THE EARTH AS A CRADLE FOR LIFE

The origin, evolution and future of the environment

T0349835

THE EARTH AS A CRADLE FOR LIFE

The origin, evolution and future of the environment

Frank D. Stacey
Jane H. Hodgkinson

CSIRO Earth Science and Resource Engineering, Australia

World Scientific

NEW JERSEY · LONDON · SINGAPORE · BEIJING · SHANGHAI · HONG KONG · TAIPEI · CHENNAI

Published by

World Scientific Publishing Co. Pte. Ltd.

5 Toh Tuck Link, Singapore 596224

USA office: 27 Warren Street, Suite 401-402, Hackensack, NJ 07601

UK office: 57 Shelton Street, Covent Garden, London WC2H 9HE

Library of Congress Cataloging-in-Publication Data
Stacey, F. D. (Frank D.)
     The earth as a cradle for life / Frank D. Stacey (CSIRO, Australia) &
Jane H. Hodgkinson (CSIRO, Australia).
       pages cm
     Includes bibliographical references and index.
     ISBN 978-9814508322 (alk. paper)
     1. Earth sciences--Textbooks. 2. Environmental sciences--Textbooks.
I. Hodgkinson, Jane H.   II. Title.
     QE26.3.S73 2013
     550--dc23
                                                            2013007925

British Library Cataloguing-in-Publication Data
A catalogue record for this book is available from the British Library.

Cover photo: The northen part of the western hemisphere of the Earth, compiled from NASA and NOAA data by Reto Stockli, Alan Nelson and Fritz Hasler. Photo courtesy of NASA.

In-house Editor: Amanda Yun

Printed in Singapore by World Scientific Printers.

Photo by K.-D. Keller of a bust in the Royal Greenwich Observatory.

"...if I shall seem to advance any thing that looks like Extravagant or Romantick the reader is desir'd to suspend his Censure till he have considered the force and number of the many Arguments which concur to make good so new and so bold a Supposition", Edmund Halley (of comet fame), *Miscellanea Curiosa,* Vol. 1, p. 44 (1705).

Preface

The primary purpose of our essay is to step back from the immediate problems that are addressed in most environmental discussions and take a long term view of the Earth as the nursery in which life developed over billions of years and in which it can continue, probably, for billions more. A central component of our message is that advanced life forms, including ourselves, are not restricted to a limited future by anything fundamental about the Earth or its astronomical environs. Hiccups in the basic components of the environment will occur from time to time, but life is resilient and will survive them. The environmental problems of our own making are fundamentally different from the natural ones and an appeal to the resilience of life to overcome them is dangerous. Rather, we should be viewing the Earth as a basic life support system and maximising its value by assuming that it will last forever and using it accordingly.

If you think that it would be possible, in principle, to design a planet better suited to our existence than the one we have, think again. We are products of our environment, matched to it by many millions of years of evolutionary trial and error. To understand our environmental needs at a fundamental level we must study the Earth not only as it is but also, as far as possible, its origin and history, because the evolution of biological life was closely linked to the evolution of the Earth. In trying to understand what makes the Earth suitable for our existence, it is helpful to compare it with other planets, but, being restricted to the solar system, there is a very limited number of planets available for comparison and they are all very different. Features that make the Earth a special place include the following:

- It is a solid planet with a mass large enough for its gravity to hold an atmosphere of the heavier gases, nitrogen, oxygen, and argon, but not so large that it also retains the light gases, hydrogen and helium.

- In a universe dominated by hydrogen, it has an atmosphere rich in oxygen.
- It has both abundant surface water and large areas of land.
- In spite of evidence for a progressive increase in the output of the Sun, the Earth's surface temperature has been steady enough to allow water to exist in all three phases (solid, liquid and gas) for more than 4 billion years.
- It has a magnetic field strong enough to protect the atmosphere from the solar wind (the flow of energetic particles from the Sun) and low energy cosmic rays.
- Its rotation is fast enough to give moderate day-night temperature variations and is misaligned with the orbit sufficiently to ensure that all latitudes receive some sunlight and do not indefinitely accumulate water or carbon dioxide that would freeze out at the low temperatures of permanent darkness.
- The Earth's Moon is the only large satellite in the inner solar system. It has important environmental effects and may have more influence than is yet understood.

This partial list gives a glimpse of the features that have made the Earth a compelling topic of investigation for generations of scientists. But, now that we are alert to the impact of human activity, the need for a fundamental understanding assumes a new urgency. This understanding must not be restricted to professional scientists, but must be shared by the wider community and political decision-makers in particular. Translating what is known into widely accessible forms is a multi-stage process requiring the skills not only of scientists, but of journalists and the media. Our aim is to take a step in this direction by clarifying what is known about the evolution of the Earth and natural controls on the environment. We conclude that current human-induced effects cannot be accommodated by the natural controls, even over millions of years, and that major policy changes are called for. The necessary political and social decisions must not be evaded by supposing that they can wait for more information.

In our list of contents we identify three general subject areas roughly corresponding to the three components of our subtitle. The first four chapters concern the origin and underlying physical framework of the Earth, the essential starting point of the environment. The following eight chapters discuss aspects of its evolution to the present state — a process that took more than 4 billion years. These chapters emphasise that the Earth's environment is a result of prolonged interactions between its component parts and with the rest of the solar system. In the course of its development it has supported life for at least 2 billion years (much more by some estimates) but on the available evidence it is, in this respect, unique in the solar system. At several points we comment on the wide diversity of environments in which life is found and the corresponding diversity of life forms, which call for an explanation of the Earth's biological uniqueness. Over a long enough time scale, the evolutionary adaptability of life allows it to adjust, not just to widely differing environments but to the changes in them that drive biological evolution. Mars once had surface water, which was the starting point for life on Earth. Did it once have primitive life that was snuffed out by the almost complete loss of water? And is the desertification so much more extreme than anywhere on Earth that no life could survive, or was the life-friendly environment too brief to allow the necessary evolutionary adaption? These are questions still under investigation that bear directly on the unique habitability of the Earth.

Acknowledging that the Earth continues to evolve, we conclude that it has everything in place to provide for life, not just for millions, but billions of years more. Can we really understand the significance of our tampering with the environment without being aware of how it reached the present state? We believe not. The first 12 chapters summarise what has been learned, to provide the knowledge from which to assess the material presented in the final chapters. These concern the environmental discussions that attract most attention, fossil fuel burning, atmospheric carbon dioxide and global warming, prompting us to reword our question: Are we demonstrating recognition of the enduring support for life that the Earth offers and treating it accordingly? Again, we believe not. We are in the process of producing irreversible changes and threatening to bequeath to future generations an Earth very different from the one we inherited.

The use of energy provides a focus for discussion of the environmental effects of human activity. It continues to increase rapidly, driven by the life expectations of an expanding population, and is a global geophysical phenomenon. The average rate of human use of energy now exceeds 16 terawatts (16×10^{12} W, 16 billion kilowatts), more than 2 kilowatts (48 kilowatt-hours per day) per person. The terawatt is a convenient unit for representing the rates of energy dissipation by many geophysical effects and we can see the global significance of this number by comparing it with the dissipations by some natural phenomena:

- flow of the world's rivers, 6 terawatts
- tides, 3.9 terawatts
- breaking waves, 5 terawatts
- heat from the Earth's interior, 44 terawatts.

The comparison draws attention to the limited range of the renewable energy sources that are, in principle, adequate for consideration as alternatives to fossil fuels.

Although the energy problem looms large and will be a major challenge in the present century, in the much longer term it is probably the least daunting of the resource concerns. Solar energy and wind power are abundantly available, in principle, and will remain so; the challenge is to harness them on a sufficient scale. But minerals extracted from the Earth are products of complex interactions that have occurred on and in the Earth over millions or billions of years. In many cases they are remarkable concentrations of certain chemical elements and compounds that will never be repeated. The prospect of exhaustion compels us to recognise that the expanding resource use by the present style of industrial activity is an extremely brief transient when viewed on a geological time scale. The popular expressions 'sustainable development' and 'sustainable mining' are oxymorons, refusals to recognise that the Earth and its resources are finite.

It is sometimes difficult to recognise which of the many statements about the environment are well informed, but the scientific principles are well understood. We approach the subject from the perspective of global geophysics and some background is assumed in many of the ideas discussed. However, we have aimed to present the basic observations and the theories connecting them in a way that is comprehensible and satisfying for readers who do not have access to the professional literature and would find mathematical arguments an unwelcome distraction, but want sufficient background to make their own judgements. The range of concepts is diverse and the chapters are divided into short sections, with section headings, to assist in navigating through them and in choosing sections for selective reading. Some readers may want more detail, quantitative support for some of our statements and leads to important literature. For them a set of brief notes, identified by numbers in square brackets in the text, is appended. The notes help also in emphasising the significance of observations that have been neglected, for various reasons, but can be brought together to give a more coherent account of the environment than is possible without them, leading to conclusions that are summarised in a brief final chapter.

We sought comments on a draft manuscript from colleagues with a wide range of expertise and thank them for enthusiastic support and helpful advice: Habib Alehossein, Stefan Balliarno, Paul Davis, Arnold Dix, Marc Elmouttie, David Hodgkinson, Jonathan Hodgkinson, Mark Jacobson, Anna Littleboy, Michael McKillop, Bruce Peterson, Henry Pollack, Conrad Stacey, Mark Stacey, our cover artist, Chin Choon Ng, and our editor, Amanda Yun. We hope that, in this final version, they will recognise the effect of their advice.

Frank Stacey
Jane Hodgkinson

Contents

PHYSICAL AND
ASTRONOMICAL
FOUNDATIONS

1. "The age of the Earth as an abode fitted for life" (Lord Kelvin, 1899)

Age estimates without radioactivity: history of a paradox

This chapter re-uses the title of a paper by Kelvin [1.1]*, who was addressing the question: For how long has the Earth been able to support biological life? He argued that since the radiant energy of the Sun was essential to the habitability of the Earth, terrestrial life could not have existed for a period longer than allowed by the available solar energy. Several decades earlier, first a German physicist-mathematician, Hermann von Helmholtz, and then Kelvin had published papers on the age of the Sun, meaning the time for which its energy could have maintained its radiation at the present strength. Their calculations relied on the gravitational energy of the formation of the Sun. The estimated age, about 20 million years, confounded the geologists of the time because sedimentary layers in the Earth's crust required hundreds of millions of years to accumulate and they could have formed only while the Sun warmed the Earth sufficiently to maintain abundant liquid surface water, with evaporation and rainfall. Although the age estimate can be revised upwards by allowing for the central concentration of the Sun's mass, the revision is inadequate to avoid the conflicting evidence. It presented a paradox that was expressed succinctly by Clarence King, director of the US Geological Survey, in an 1893 article [1.2]: "…the age assigned to the sun by Helmholtz and Kelvin (15 million or 20 million years) communicated a shock from which geologists have never recovered". Resolution of the paradox began only after 1896, when the

---

*Numbers in square brackets identify explanatory notes at the end of this essay

French physicist, Henri Becquerel, finally announced the discovery of radioactivity, after 20 years' investigation of anomalous radiation from certain minerals [1.3]. Radioactivity as a source of heat was soon recognised and this introduced the previously unknown concept of nuclear energy, reopening the debate about the ages of both the Sun and Earth, to the relief of the geological community, although Kelvin was not easily convinced. Table 1.1 lists the radioactive elements relevant to the present discussion.

In 1863, a year after his age-of-the-Sun paper, Kelvin published an article on the age of the Earth as a habitable body, based on a quite different calculation: the rate at which it is losing heat. By then there were mines of sufficient depths to give a reliable estimate of the variation of temperature with depth, about 20°C per kilometre. By assuming that the Earth had originally been entirely molten and had formed a solid crust that progressively thickened as heat was radiated away, Kelvin showed that the increase in temperature with depth suggested a crustal thickness of about 50 km and that it would have developed in about 20 million years. Since the Earth could not have been habitable until a solid crust had formed, this was an independent estimate of the maximum duration of its habitability. The Earth cooling calculation is more often referred to than the solar energy papers, but was not as soundly based and, taken alone, would not have been too seriously regarded. However, taken together, the solar and terrestrial calculations made a forceful argument for the 20 million year age limit. Extending the above quotation from King [1.2]: "...the concordance of results between the ages of the sun and earth certainly strengthens the physical case and throws the burden of proof upon those who hold to the vaguely vast age derived from sedimentary geology". It is evident that some new physics was needed to resolve the paradox and the need was so compelling that the 1896 discovery of radioactivity was quickly hailed as that new physics, although it fell far short of a complete explanation. In the words of Lord Rutherford and a colleague, Frederick Soddy, writing in 1903 [1.4]: "The maintenance of solar energy, for example, no longer presents any fundamental difficulty if the internal energy of component elements is considered to be available, i.e. if processes of sub-atomic change are going on".

Table 1.1. Radioactive isotopes of particular environmental interest.

Element	Isotope	Half life	End product	Origin	Where found
Beryllium	^{10}Be	1.39 million years	Boron, ^{10}B	Cosmic ray bombard-ment of upper atmosphere	Marine sediment, ice cores, fresh andesite
Aluminium	^{26}Al	0.72 million years	Magnesium, ^{26}Mg	Pre-solar supernova	Extinct. Identified by Mg in meteorites
Potassium	^{40}K	1.25 billion years	Argon, ^{40}Ar, 10.5% Calcium, ^{40}Ca, 89%		Common minor ingredient of rocks. 0.0117% of potassium
Iodine	^{129}I	16 million years	Xenon, ^{129}Xe		Extinct. Identified by xenon in meteorites
Hafnium	^{182}Hf	9 million years	Tungsten, ^{182}W		Extinct. Identified by tungsten isotope ratios in rocks and meteorites
Thorium	^{232}Th	14 billion years	Lead, ^{208}Pb		Common minor ingredient of rocks
Uranium	^{235}U	0.704 billion years. Prompt fission with neutron bombardment	Lead, ^{207}Pb		Common minor ingredient of rocks
	^{238}U	4.47 billion years	Lead, ^{206}Pb		
Plutonium	^{239}Pu	24 100 years. Prompt fission with neutron bombardment	Lead, ^{207}Pb, via ^{235}U	Nuclear industry	No natural occurrence
	^{244}Pu	81 million years. 0.3% spontaneous fission	Lead, ^{208}Pb via ^{232}Th Fission products 0.3%	Pre-solar supernova	Extinct. Identified by fission xenon in meteorites

The paradox resolved but with continued difficulties

Although the discovery of radioactivity had a profound effect on Earth science as well as physics, as Kelvin recognised, it did not immediately resolve the solar energy problem. Certainly the radioactivity of uranium was not directly relevant, because, even with the completely unrealistic assumption that the Sun was composed of 100% uranium, the energy output of its radioactivity would be only half the observed radiant energy. The solar energy source was not understood until the 1930s (30 years after Kelvin's death), when fusion reactions were recognised — in particular the fusion of hydrogen to helium, which is far more energetic than any radioactive process (Fig. 1.1) and is now identified as the source of the Sun's energy. The abundance of hydrogen in the universe generally, and in the Sun and other stars (see Table 5.3), finally removed doubt about the adequacy of nuclear energy to power the Sun but, until then, its great age had to be taken on trust because it was required to account for geological events hundreds of millions of years ago.

Recognition of the heat produced by radioactivity was deemed to release geology from the restricted age limit, but consequential changes in the understanding of the Earth came slowly and Kelvin's approach to the Earth's heat was not significantly modified for another 60 years. Measurements of radioactivity and of the heat released by it in common rocks, notably by R. J. Strutt, showed that it was much stronger than needed in the Earth as a whole to explain the Earth's heat. Writing in 1906, Strutt noted that a 10 km or 20 km layer of granite could explain the entire heat flux from the Earth and conjectured that radioactivity was confined to a thin crust, with the bulk of the Earth thermally passive [1.5]. The word 'crust' thereby acquired a new meaning. It was no longer a solid layer on an otherwise molten earth, as originally envisaged by Kelvin. With its new meaning the crust was identified as a compositionally distinct layer, a veneer, on an Earth that was solid to great depth. With so much heat from a shallow source on an apparently rigid Earth, there appeared to be no reason to reject Kelvin's view of the Earth cooling solely by thermal diffusion, if at all. Deep convection of heat, which we now recognise as the driver of geological processes (plate

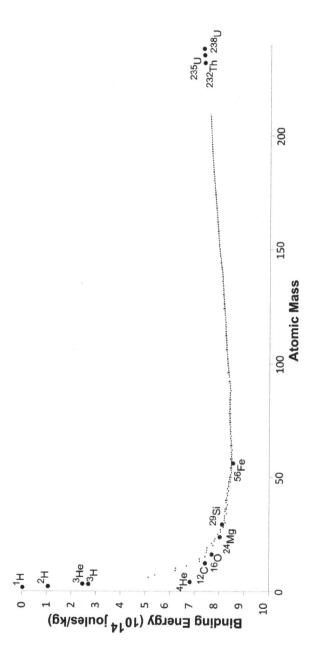

Fig. 1.1. Binding energies of naturally occurring atomic nuclei, plotted as energy per kilogram of nuclear matter, relative to hydrogen (at the top left corner). The curve is inverted to appear as potential energy, so that nuclear processes can be viewed as the tendency for everything to fall towards the lowest point, marked by iron 56. Energy is released when a uranium atom fissions, breaking into smaller pieces, or when hydrogen fuses into larger nuclei, especially helium 4. The important abundant light nuclei, helium 4, carbon 12 and oxygen 16 lie well below the trend of the curve and are particularly stable. It is obvious that the energy released by fusion of hydrogen to helium 4, a process occurring in the Sun, greatly exceeds the energy that can be derived from uranium. Isotopes relevant to energy considerations are highlighted.

tectonics, earthquakes, etc.) had been postulated in the late 1800s as a way to circumvent Kelvin's age constraint, but was deemed irrelevant because of the solar energy problem. So, although radioactivity was quickly accepted as a terrestrial heat source, the fact that it was found to be concentrated in a shallow crust removed the perceived need for deep convection in the Earth, which was not widely accepted until about 1960.

Dating by radioactivity

Rutherford was quick to recognise the possibility of using radioactivity to date geological events. His original method was to measure the accumulation of helium in uranium-bearing rocks. He did not know, at that time, that uranium has two naturally occurring isotopes, being a mixture of atoms with different masses and different rates of radioactive decay. They are identified by their masses, 238 and 235 atomic mass units. The heavier isotope, which is given the symbol ^{238}U, comprises more than 99% of total uranium. The lighter isotope, ^{235}U, is more strongly radioactive and so is shorter-lived and now less abundant. Both decay to isotopes of lead in several stages, emitting alpha particles (helium nuclei) as they do so (Table 1.1). Helium (He) is a chemically inert gas that is retained by solid materials, such as rocks, if they remain cool, but diffuses out if they are warmed. The helium found in rocks and minerals has been produced in situ by the decay of uranium (and also thorium, if present) and increases with time. In spite of the complications that were not all understood, Rutherford had a reasonable estimate of the rate at which uranium decays and the amount of helium that its decay produces. He used the He/U ratios that he measured in minerals to calculate the times required for the He to accumulate and the numbers he came up with were hundreds of millions of years, even exceeding the ages that sedimentary geologists had called for. In fact the method generally underestimates the formation ages of rocks, because few of them have remained sufficiently cool to retain all of the helium, and a better method, using the decay of uranium to lead, was soon applied by B. Boltwood, an American colleague of Rutherford.

In the century since Rutherford's pioneering measurements, the use of radioactive decays to determine the ages of geological materials (termed radiometric dating) has become very sophisticated, with instruments that measure precisely the abundances of isotopes that may exist only in trace amounts. The uranium-helium dating method is now used only to investigate the thermal histories of rocks that are dated by other methods, but the decay of uranium to lead is quite widely used. Measurements using the mineral zircon (zirconium silicate) have particular advantages. Uranium and lead diffuse in it less readily than in most other minerals and the zircon crystal structure accepts substitution of uranium for zirconium but almost completely rejects lead. This means that the lead found in zircons is almost entirely radiogenic (produced by radioactive decay), requiring little allowance for *initial lead*, incorporated when they formed. In this situation, the age determination is both reliable and relatively simple because it requires only the measurement of a ratio of the lead (Pb) isotopes, $^{207}Pb/^{206}Pb$, that are produced by the decays of the two isotopes of uranium. The present ratio of uranium isotopes is found always to be the same, but ^{235}U decays much faster than ^{238}U and is now only 0.72% of total uranium, but was 24% when the Earth first formed [1.6]. Thus, zircons of great ages accumulated ^{207}Pb (from the shorter lived ^{235}U) much faster in their youth and now have higher $^{207}Pb/^{206}Pb$ ratios than those in younger rocks.

In general the initial lead cannot be neglected and dating by lead isotopes requires multiple samples of the same age but different ratios of uranium to lead, such as the various minerals in a single rock. When an igneous rock is formed from melt, the isotopes have been homogenized, so that all of the minerals start with the same isotopic ratios, which then evolve by additions of ^{207}Pb and ^{206}Pb according to their uranium contents. Since the uranium isotopic ratio is the same in all of them, the ratio of the added ^{207}Pb and ^{206}Pb (termed radiogenic lead) is also the same and, if this could be measured independently of the initial lead, included when the minerals formed, it would provide a measure of their age. A graph of ^{207}Pb vs ^{206}Pb for several minerals with different uranium abundances would be a straight line, with a gradient controlled by the age. Such a simple measurement is frustrated by the initial lead, introducing a complication, which is resolved by measurement of another lead isotope, ^{204}Pb, which is not a product of any natural decay process and serves as a reference standard. The result is a linear relationship between the ratios ^{207}Pb/^{204}Pb and ^{206}Pb/^{204}Pb [1.7], referred to as an isochron (equal time line), displayed as a graph with all samples of the same age joined by a straight line, as in Fig. 1.2. When the fit is a good one, the data are said to be concordant, indicating reliability of the age estimate. Discordant data arise when a disturbance, such as heating, has occurred at some stage, causing diffusion of isotopes from or between minerals. Attempts to infer ages from discordant data are fraught with difficulty and rarely successful.

Ages of the solar system and the heavy elements in it

Figure 1.2 is a particularly important lead-lead isochron, obtained from measurements on meteorites. They are clearly concordant, with a well-determined common age. From the plotted data this is, with the uncertainty inferred from the scatter of the data, 4.54 ± 0.03 billion years. The meteorites are identified as fragments of asteroids. An interesting piece of evidence for this is a plot of the orbits from which meteorites were observed to fall. They all extend into the asteroidal belt (Fig. 1.3). Most of the data for this figure were obtained from trails in the upper atmosphere that were photographed with multiple, well separated cameras fitted with timing shutters. With the exception of a few special types that are fragments knocked off the Moon or Mars by major impacts, the meteorites were never parts of planetary sized bodies and they cooled quickly once they were formed. The isochron dates their formation, with little scope for differing ages or later processing. The planets, including the Earth, have had more complicated histories and there are no terrestrial materials that have survived unaltered since that time, but an indicative data point for the Earth as a whole was obtained by averaging data for sediments from all of the oceans. This is plotted as the cross in Fig. 1.2. The near coincidence of the marine sediment data point and the meteorite isochron is an important piece of evidence that the Earth and meteoritic bodies (asteroids) formed at the same time from the same mixture of elements and isotopes, but with the implicit assumption that the sediments, and therefore the continents from which they were eroded, provide an unbiased average of the isotopic ratios in the Earth as a whole.

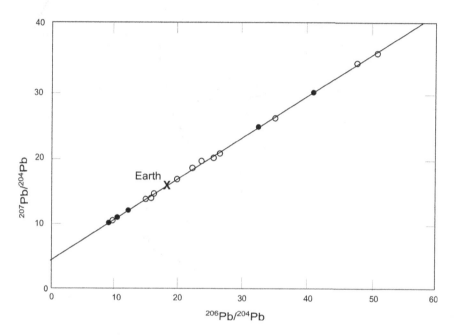

Fig. 1.2. The lead–lead isochron for meteorites, which dates the formation of the solar system. The open and solid circles are data points from two laboratories, and each represents a different meteorite. The global average of data from marine sediments is plotted as the 'Earth' point, indicating a common origin for the Earth and the asteroids, of which the meteorites are fragments. The data extend well beyond the range of this graph.

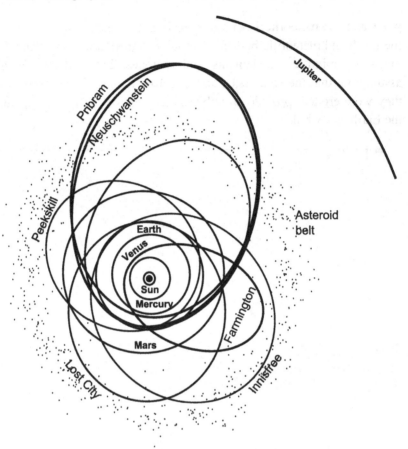

Fig. 1.3. Pre-arrival orbits of seven meteorites from plots by McSween [1.8] and Spurný et al [1.9]. All of them extend from the asteroidal belt between Mars and Jupiter.

Some radioactive isotopes are produced in the atmosphere by cosmic ray bombardment or arrive with interplanetary dust that has had prolonged exposure to cosmic rays. Relative to the age of the Earth, they are all short lived. But there are 20 longer-lived isotopes that were incorporated in the Earth and meteoritic bodies when they formed and still exist in measurable quantities [1.10]. ^{235}U is the shortest lived of the latter group, with a half-life of 0.7038 billion years, meaning that its abundance decays to half of an initial value in that time, a quarter in twice that time, and so on. The meteorite isochron age of 4.54 billion years is 6.45 times this half-life, so that, initially, ^{235}U was $2^{6.45} = 87$ times as abundant as at present, as in Note [1.6]. These numbers invite an important inference. The absence in the Earth of isotopes with half-lives much shorter than a billion years cannot be explained by supposing that only long-lived ones were produced. From what we know about synthesis of the elements, not only would many shorter-lived species have been produced at the same time, but they would have included many light nuclei, which would have been more abundant. (The distribution of elements is strongly biased to light species). The fact that short-lived radioactive elements no longer exist in observable quantities (except for those identified as products of cosmic ray bombardment) is consistent with the 4.5 billion year age of the solar system. But the synthesis of the heavy elements could not have occurred very long before the formation of the planets, because that would have required an implausibly large initial abundance of ^{235}U. As Rutherford first noted, the origin of the radioactive elements can be dated [1.11]; from what we now know about the age of the solar system, it is evident that the interval between synthesis of these elements and the accretion of the planets was much shorter than the subsequent life of the planets.

The existence in meteorites of *orphans*, that is, isotopes that are products of radioactive decays of now extinct parents, allowed the synthesis-accretion interval to be investigated more closely. Xenon isotopes produced by the decay of an iodine isotope, ^{129}I, and by spontaneous fission of an isotope of plutonium, ^{244}Pu, (not to be confused with ^{239}Pu that is used in the nuclear industry) have attracted particular attention. The decay of a short-lived isotope of aluminium, ^{26}Al, (half-life 0.72 million years) to magnesium, ^{26}Mg, is arguably more useful because these are abundant elements. The short half-life of ^{26}Al, means that the identification in aluminium-bearing minerals in meteorites of ^{26}Mg, without corresponding abundances of the more common magnesium isotopes, means that the ^{26}Mg was formed in situ by decay of ^{26}Al, some

William Thomson, Lord Kelvin, 1824–1907
British physicist and engineer, most widely known for his work on thermodynamics and the absolute temperature scale (degrees Kelvin), but with major contributions to several fields including electromagnetism, telegraphy, marine cables, and ships' compasses. For 50 years he was Professor of Natural Philosophy at the University of Glasgow, where his father had been Professor of Mathematics. Kelvin's calculations on the ages of the Sun and Earth, which pre-dated recognition of nuclear processes, confounded geologists in the late 19th century.

Ernest Rutherford, Baron Rutherford of Nelson,
1871–1937
New Zealand-born physicist with extensive pioneering discoveries in atomic and nuclear physics, first at McGill University (Canada), then the University of Manchester (UK) and finally Cambridge. His measurements of radioactivity led to the first realistic estimates of the ages of rocks and the time of origin of uranium, releasing geological studies from the constraint of Kelvin's estimate of the Earth's age.

of which survived until the minerals had formed. The interval between synthesis of the radioactive elements and accretion of the solid grains that accumulated as meteoritic parent bodies was no more than a few times the half-life of ^{26}Al (0.72 million years), compared with the subsequent 4.5 billion year life of the solar system. The ^{26}Al was evidently produced at the same time as the very heavy elements, thorium, uranium and plutonium, that could only have been produced in the neutron-rich flash of a supernova. Their existence means that a supernova preceded the formation of the solar system by no more than a few million years, with debris from it mixing into a cloud of gas and dust that had been accumulating from other sources over billions of years. The obvious inference is that the supernova was directly responsible for collapse of the cloud to produce the solar system and the simplest explanation is that the radioactive supernova debris ionized the cloud, making it an electrical conductor, with highly mobile electrons. Complex turbulent motion of a fluid electrical conductor has two effects. By moving in a magnetic field, it generates electric currents, and the currents produce a magnetic field. With suitable motion, the currents and magnetic field are mutually reinforcing, as in the Earth's core, where the Earth's field originates (Chapter 4), and the combination produces a force that opposes the motion, which continues only if it is maintained by independent forces. The turbulent motion of the solar nebula would have been damped in this way, reducing it to a disc shape, from which the solar system accreted. This happened only when the nebula had become an electrical conductor with the arrival of fiercely radioactive supernova debris.

Early development of the basic structure of the Earth

In the spirit of Kelvin's original enquiry, the date of formation of the Earth is only part of the story. When did it become habitable? The starting point was the establishment of its broad structure, with the core, mantle, crust and oceans. We consider first the core, the dense central part of the Earth, with a radius of 3480 km — 55% of the radius of the Earth. Its density is explicable only if it is composed largely of metallic iron, which is known from the iron meteorites and its abundance in the Sun, to be quite common in the solar system. Most of the core is liquid, with a density somewhat lower than that of iron-nickel alloy, as found in iron meteorites, so it is identified as a molten alloy with some light solutes. But there is also a group of heavier elements that are termed siderophiles (iron-lovers) that would have dissolved in the core. One of them is tungsten, represented by the symbol W (German: wolfram). A particular tungsten isotope, ^{182}W, provides evidence that the core separated from the rest of the Earth very early, because at least some of it is a radioactive decay product of an extinct isotope of hafnium, ^{182}Hf, which has a half-life of 9 million years. Hafnium is not a siderophile element and remained with the silicate (rocky) parts of the Earth, the mantle and crust, when the core separated, taking with it much of the Earth's tungsten. The hafnium decay that occurred after core separation enriched the mantle in ^{182}W, without a corresponding increase in the other tungsten isotopes. All terrestrial samples (derived from the crust and mantle because the core is inaccessible) are systematically richer in ^{182}W than the meteorites which originated from the same isotopic mix. This means that the core formed before all the ^{182}Hf had decayed, that is, not much more than about 9 million years after the origin of the heavy elements. It now appears inevitable that the core formed during accretion of the Earth, not as a separate event, and that it took with it much of the Earth's tungsten, but with less ^{182}W than is now in the meteorites, in which it has accumulated since by the decay of ^{182}Hf. We have no way of sampling the core, but can infer what happened because the mantle retained all of the terrestrial ^{182}Hf, and so has accumulated a higher proportion of ^{182}W than the meteorites.

Now we consider the separation of the continental crust from the mantle. The mineral zircon, which we referred to in connection with lead-uranium dating, provides an important clue. It occurs in what are referred to as acid rocks, meaning those that are rich in silica, which, in its pure form, is quartz. Granite is a common rock of this type. These are the rocks that make up the bulk of the continents and are not found on the ocean floors. That zircons have been found with ages exceeding 4 billion years (4.4 billion years in the extreme case) is evidence that continents formed very early. That carries a further inference that is crucial to our understanding of the early development of the Earth's structure: there were oceans also at that time. Most of the Earth, that is the mantle, is composed of what is termed ultrabasic rock, meaning very low in silica, and the separation of acid rocks from it is a two stage, or even three stage, volcanic process, driven by convection of the mantle. The first stage is the production of basalt, a basic (less than ultrabasic) igneous rock that forms the ocean floor crust. It is a product of ocean floor volcanism, driven by convection in the underlying mantle, as discussed in Chapter 5. It survives, on average, for about 90 million years (but less when the Earth was young), cooling as it moves to a subduction zone, where it is carried down again into the mantle, as part of the convective cycle, taking with it some of the ocean floor sediment and sea water. As the sinking material is reheated, the sea water acts as a flux, lowering the melting point and generating new lava that is more silica-rich than the basalt, and erupts in volcanic zones, typically as andesite (named from the Andes mountains). Granite is even more silica-rich and requires a further stage, or stages, of recycling. Table 1.2 lists average compositions of these rock types. All this must have happened more than 4 billion years ago and involved sea water. Without sea water, volcanism would produce only basic igneous rock, low in silica. The silica-rich acid rocks that include zircons depend on volcanism that involves subducted sea water, as discussed in Chapter 5. Thus, the basic structure of the Earth, with proto-continents and oceans was established more than 4 billion years ago and, from the earliest known zircons, we suggest 4.4 billion years ago. By contrast, as discussed in Chapter 7, the atmosphere has changed dramatically and the nitrogen content is the only component that now matches what the Earth probably had when it first formed. Mars, which once had sufficient water, may have some acid rock, but otherwise both the Earth's continents and its atmosphere, as well as the oceans, are unique in the solar system.

Table 1.2. Major components (percentages by mass) of selected crustal igneous rocks that arise from three stages of volcanic separation from the mantle.

Stage	Rock type	silica, SiO_2	magnesia, MgO	iron oxides, FeO $+ Fe_2O_3$	alumina, Al_2O_3	lime, CaO
1	ocean ridge basalt	47.5	14.2	9.5	13.5	11.3
	plume basalt	49.4	8.4	12.4	13.9	10.3
2	andesite	59.2	3.0	6.9	17.1	7.1
3+	granite	72.9	0.5	2.5	14.5	1.4

Beginnings of life on Earth

With the very early development of the structure of the Earth, complete with continents and oceans, we can take a fresh look at the more limited time span of advanced life forms. In Kelvin's use of our adopted chapter title, he was very undemanding. His only requirement for habitability was that the Earth must have a solid crust. With his age limit removed, we can consider what else was needed. One of the requirements for development of life was water but, as we have mentioned, that was available very early, perhaps from the beginning. What does 'life' mean in this context? By including the simplest of self-reproducing organisms, it appears that the Earth has supported life for at least 2 billion years of its 4.5 billion-year existence and, by some interpretations of the fossil record, 3.8 billion years [1.12], but that is disputed [7.13]. The more advanced life-forms seen in the fossil record, and life on land, are restricted to no more than the last 600 million years, but once they began, they rapidly diversified. It would be possible, in principle, to argue that the delay was simply the time required for evolution to advance to that point, but that cannot have been the entire reason. Evolution is a response to the environment, which was changing at that time with decreasing carbon dioxide, increasing atmospheric oxygen and, with it, ozone that gave protection from ultraviolet radiation. How did it happen? Is it possible that it is the result of a balance that is so sensitive that it could be reversed, either naturally or by human activity? What other features of our environment are as essential to advanced life as we know it and how vulnerable are they? The fossil record gives some clues (Chapter 9) and it is useful also to compare our atmosphere with the atmospheres of our nearest neighbouring planets (Table 7.1).

We need to be clear about the basic point that life developed on the Earth when conditions were favourable for it. Given the necessary environment, the only requirement is sufficient time. The existence of organic molecules in remote parts of the solar system (and further afield) has been made a reason for postulating that life was 'planted' on the Earth by an injection of material from space. If these molecular structures can appear under much less favourable conditions than we have, then they can more easily appear here and may have done so in the upper atmosphere almost as soon as the Earth formed [1.13]. What this shows is that the process is inevitable with the right environmental conditions. Our task is to deduce just what these conditions are, how they arose and what further changes must be expected.

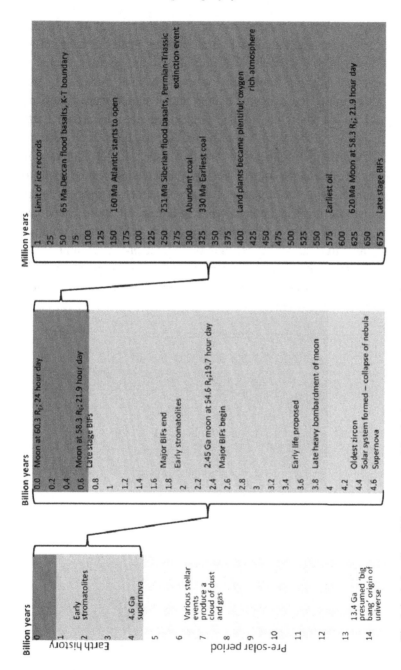

Fig. 1.4. A time line of the events resulting in the present environment. See also Table 9.1 for details of geological periods.

Figure 1.4 conveys an impression of the time that it has taken for the Earth to form and develop to the present state of habitability. Although we argue that there was nothing in the various stages of development outside the range of normal physical processes, they involved an extended sequence of events, all of which were necessary. There could be no short cut to the production of a habitable Earth. Regarding its formation as an experiment, it must have been repeated countless times in the vastness of the universe, with all the random permutations of alternative steps. We have no clear idea how many of these experiments have been successful, but the fraction must be small. That is a matter for philosophical contemplation. Of more immediate concern is the realisation that we are living in a very special situation, resulting from physical, chemical and biological processes and events that we need to understand to ensure that the environment, and the opportunities and resources that it presents, are not too seriously compromised by our activities.

2. Rotation, tides and the Moon

Uniqueness of the Moon — why are there so few satellites in the inner solar system?

The outer four planets have numerous satellites. Many of the small ones are almost certain to be captured asteroids or minor bodies, especially the distant satellites of Jupiter, but all of the gaseous giants also have large satellites that were clearly formed in or near to their present regular orbits. Although not strictly a planet, Pluto is now known to have five satellites, one of them, Charon, with a mass that is about 12% of that of the parent planet, and even some asteroids have satellites (or partners, making them twin asteroids). The formation of satellites is a normal accompaniment to planetary accretion, but in the inner solar system the four terrestrial planets (represented in Fig. 4.1) have only three satellites between them. Mars has two that are very small and irregular in shape and give the impression of being captured asteroids, while Venus and Mercury have no satellites at all. The Earth's large Moon is exceptional. Numerous theories have attempted to explain its existence with the presumption that it is anomalous. But, is it anomalous, requiring a very special explanation for the origin of the Earth-Moon system? The questions that should be asked are not 'Why does the Earth have a large moon?', but 'Why is it the only one?' and 'Why do Mercury and Venus have none?'. The simple answer to both questions is 'tidal friction' and it holds an essential clue, not just to the origin and history of the Moon, but also to the history of the Earth. Tides dissipate rotational energy and modify orbits. The Moon has been important to the development of the Earth, not least because of the tide that it raises, but its formation and subsequent history followed well established physical principles, with no

21

special conditions or circumstances. We reject a supposition that the formation of the Moon required unusual physical circumstances, such as a massive impact on the Earth.

Tides and tidal friction

The study of tides has a long history [2.1]. The spirit of enquiry in ancient Greece resulted in a quite detailed record of basic observations, but not in the Mediterranean, where tides are insignificant. The Greeks were alerted to tidal phenomena by visits to Britain and the Atlantic coast of Spain, which have strong marine tides. The first documented record appears to be by Pytheas, in about 330 BC. He noted the coincidence of the tidal cycle with the phases of the Moon, but with two tides per day and amplitude varying in a monthly cycle. His conclusion that the tides were somehow controlled by the Moon was incorporated in reports by later investigators and commentators, but not accepted by Galileo almost 2000 years later, and not satisfactorily explained until Isaac Newton presented his theory of gravity in 1687. Understanding was delayed by the complications of the marine tides and, in particular, an early 700s AD report by Bede, the English monk and scholar, of a systematic variation in the times of tides at ports down the east coast of England. Particularly relevant to our discussion is a perceptive report by an Italian scientist, Julius Caesar Scaliger, who, in 1557, suggested that Atlantic tides were driven not just by the Moon, but by a resonant oscillation of the ocean. This is considered below in connection with the history of the Earth's rotation and is referred to again in Chapter 6. The dissipation of rotational energy by tides appears to have been suggested first by Immanuel Kant in 1754, but was not followed up for another century.

The gravitational forces of the Moon and Sun each deform the Earth in the manner of Fig. 2.1. This is tidal deformation, which is obvious in the oceans but affects the solid Earth as well. Because of the proximity of the Moon, the tide raised by it is slightly more than twice as strong as the tide raised by the more massive but more distant Sun and we concentrate attention on the lunar tide. It can be considered to arise because the average gravitational force of the Moon that keeps the two

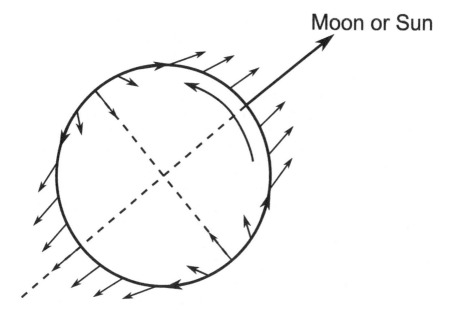

Fig. 2.1. The pattern of the tidal force on the Earth. At any point this is the difference between the gravitational force of the Moon or Sun at that point and the force required to maintain the orbital motion.

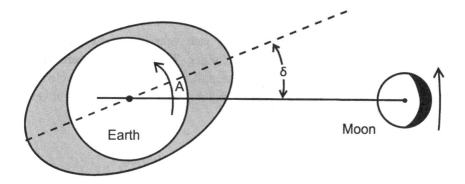

Fig. 2.2. The effect of tidal energy loss is to delay the tide. The point A was directly below the Moon 12 minutes ago and the Earth has rotated by the angle δ (2.9°) in this time. The force between the delayed tidal bulge and the Moon slows the Earth's rotation and causes the Moon to swing into a wider orbit.

bodies together in their orbital motion is the force at the Earth's centre, but the attraction to the Moon is stronger than needed on the near side and too weak on the far side, resulting in an ellipsoidal deformation with its long axis in the direction of the Moon. The orbital motion of the Moon about the Earth is slow compared with the rotation of the Earth and the deformation remains aligned with the Moon as the Earth rotates, so that each point on the Earth passes through successive tidal highs and lows. The tidal bulge is not exactly aligned with the Moon, as it would be if the Earth were perfectly elastic and, more importantly, if the ocean tide were in equilibrium with the lunar gravity. Both the solid Earth deformation and the marine tide dissipate energy, and this takes effect as a drag on the tidal motion, delaying the response to the gravitational force and shifting the tidal bulge in the manner of Fig. 2.2. But, since the bulge is caused by the Moon, the Moon 'tries' to pull the bulge back into alignment with itself, causing a tidal torque, which opposes the Earth's rotation, slowing it down. This is tidal friction. The equal and opposite torque exerted on the Moon by the tidal bulge acts in the direction that would accelerate the orbital motion and causes the Moon's orbit to expand. The effect is most precisely measured by observing the motion of the Moon, which is currently receding from the Earth by 3.7 cm/year. Allowing for the added effect of the solar tide, the corresponding slowing of the Earth's rotation causes the length of the day to increase by 24 microseconds per year — that is, for each year that passes, the day becomes 24 millionths of a second longer. This may not seem very much, but the effect was stronger in the past, when the Moon was closer, and over geological time has had a major effect on the history of the Earth and Moon.

Rotation of the Earth and evolution of the Moon's orbit

The present tidal friction is well observed and the theory is basically simple [2.2]. Extrapolation back in time gives the Earth-Moon distance varying as the broken line in Fig. 2.3. It suggests that the Moon may have been in orbit for no more than 1.6 billion years, whereas both the

Earth and Moon have been geologically stable for much longer than that. The present rate of tidal dissipation is not representative of the past. The only obvious explanation depends on changes to the configuration of continents and oceans, coupled with Scaliger's suggestion of a strong marine tidal resonance, and we pursue this in Chapter 6. But there is no possibility of determining ocean floor geometry in the remote past in sufficient detail for a quantitative extrapolation of tidal friction and we appeal to geological evidence of the rotation rate. In a simple way, we can see qualitatively what to expect. Gravity depends on the inverse square of distance ($1/r^2$) and tidal force as the inverse cube ($1/r^3$), because it is caused by the gravity gradient, that is the gravity difference across the Earth. Tidal energy is proportional to the square of amplitude ($1/r^6$) and so is the rate of dissipation of this energy. Complications have arisen from variations in ocean geometry, but their effect is outweighed by the strong dependence of tidal dissipation on the distance, r, between the Earth and Moon. The evolution of the lunar orbit was much faster when the Moon was closer.

The first measurements of tidal periods in the remote past used growth increments in the fossilised shells of marine creatures that had grown in tidal zones and developed layered structures that indicated tidal and seasonal cycles. They have been termed 'paleontological clocks'. The conclusion clearly emerged that the rotation of the Earth and the lunar orbital motion were both faster a few hundred million years ago, but not by as much as the extrapolation used to calculate the broken line in Fig. 2.3 suggests. Subsequently it became evident that data obtained from the layering in sedimentary rocks that had accumulated in tidal estuaries are more reliable and extend over a much wider time scale. It is useful to consider how observations of sedimentary layers lead to estimates of the tidal periods because there is obviously no clock built into the rocks that would allow the periods to be observed directly. The period of the Earth's rotation relative to the Sun (the solar day) and the period of the Moon's orbital motion (the lunar month) are both progressively increased by tidal braking and the lunar tide is driven by the difference between them. However, the period of the orbit about the Sun (the year) is not affected because in that case the energy of the orbital motion completely dwarfs the tidal energy and so the solar tide is controlled solely by the length of the solar day. The difference between this and the period of the lunar tide is therefore a measure of the orbital period of the Moon. The two tidal periods are nearly equal and the combination is seen as a beat, a cyclic variation in the amplitude of the tide as the solar and lunar tides either add or subtract (Fig. 2.4). The beat period measures the difference between the two tidal periods. If an annual cycle is also discernible it serves as a cross-check, because the length of the year is unaffected by tidal friction. Greatest interest attaches to the Earth-Moon distance, as in Fig. 2.3, and this follows directly from the lunar orbital period. 'Tidal rhythmites', sediments in which tidal cycles have resulted in well-developed layering, have been used most effectively by G. E. Williams [2.3], whose data at 620 million years (a) and 2.45 billion years (b) are joined by the solid line in Fig. 2.3.

As we demonstrate in Chapter 6, the 620 million-year and 2.45 billion-year data points show that, in this time interval, tidal friction was only a third as strong as the simple extrapolation from present day observations (the broken line in Fig. 2.3) suggests. The data for later times from the paleontological clocks confirm that the dramatic increase in tidal friction is a geologically recent phenomenon. The explanation (Chapter 6) needs to be convincing because evolution of the lunar orbit is critical to understanding the early history of the Earth and to implications of the 3.8 billion year data point (c) in Fig. 2.3, which arises from a different consideration that we discuss presently.

When tidal friction runs its full course, it completely stops the rotation of a planet or satellite, relative to another body that raises a tide in it. In this respect tidal friction resembles ordinary mechanical friction between bodies, completely stopping relative motion that is not maintained by an external force. As is obvious in the case of friction of the solid tide, but applies also to the marine tide, the lag angle, δ in Fig. 2.2, does not diminish as the relative rotation slows, but maintains the tidal torque until the relative rotation stops. The familiar example is the Moon, which presents a constant face to the Earth. The amplitude of the tide raised in one body, mass M (the Earth) by another of mass m (the Moon) is proportional to the ratio of the masses, m/M. Conversely,

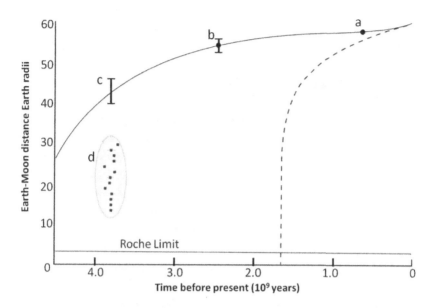

Fig. 2.3. Extrapolation of the Earth-Moon distance to the remote past. The broken line is a simple extrapolation of present day observations of the tidal lag. Data points at 620 million years (a) and 2.45 billion years (b) were obtained by Williams [2.3] from sedimentary rocks showing layering with tidal periodicities. The 3.8 billion year point (c) is an estimate for the time of a bombardment of the Moon that caused the major craters, that are dated by the lunar impact melts (d), as in Fig. 2.5. The big difference between tidal friction now and in the remote past has been a paradox that we discuss in Chapter 6.

the tide raised in the Moon by the Earth is proportional to M/m and is, therefore, $(M/m)^2$ times the tide raised in the Earth. Since $(M/m) = 81.3$, the tidal deformation of the Moon and corresponding tidal friction were very large and would have caused the Moon to present a fixed face to the Earth within a few million years of its formation. This state is maintained as the orbit evolves. As another example of this tidal end point, Pluto and its large satellite Charon both present fixed faces to each other. If the Moon were appreciably bigger than it is (and not much more distant), then the Earth would now present a fixed face to it, rotating with a month long day.

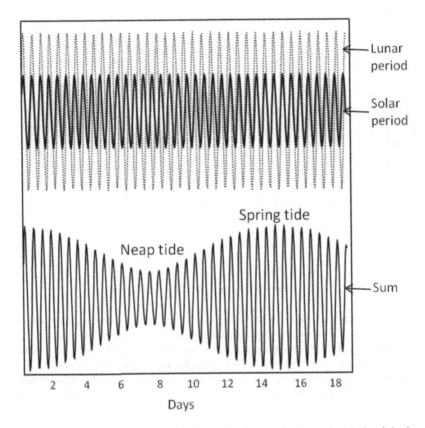

Fig. 2.4. Superposition of two oscillations with the amplitudes and periods of the lunar and solar tides. The beat period, the interval between successive spring or neap tides, the maxima and minima of the tidal amplitude, is a measure of the difference between the two component periods.

Venus and Mercury could not now have satellites, even if they once did so

Now we can consider what this means for Mercury and Venus, both of which are closer to the Sun than is the Earth and therefore have stronger solar tides. Since they lack oceans, their tidal friction arises only from the imperfect elasticity of rocks. In neither case has the axial rotation been completely stopped by tidal friction but both are very slow. It takes 59

Earth days for Mercury to rotate once about its axis, a time that is in resonance with its orbital period; Venus takes 243 Earth days to rotate once, and its rotation is opposite in direction to its orbital motion. In the present context the important point is that the rotations are so slow that any satellites would be orbiting faster than the planetary rotations. In this situation the effect of tidal friction is opposite to what we see in the Earth-Moon system, in which the Moon's motion is very slow compared with the Earth's rotation. The friction opposes the relative motion, tending to spin-up very slowly rotating planets and retard the motions of their satellites, causing them to spiral inwards towards the planets and merge with them. This can be understood by referring to Fig. 2.2, which shows the Earth's tidal bulge ahead of the position of the Moon, so that the Moon's gravity is pulling the bulge back towards alignment with itself. The gravitational effect of the bulge on the Moon is to apply an accelerating force in the direction of its motion, causing its orbit to expand. In the case of a satellite orbiting faster than the planetary rotation, the tidal lag causes the planetary tidal bulge to be delayed, relative to the position of the satellite, imposing a retarding force on it and causing the orbit to shrink. This would have happened to any satellites of Venus or Mercury. It is improbable that they never had satellites, but they could not do so now because tidal friction would have destroyed them. This process is seen to be occurring at the present time with Phobos, the very close inner satellite of Mars, which orbits three times per Martian day. In spite of the small mass of Phobos, the friction of the tide which it raises in Mars suffices to cause a measurable rate of decrease in its orbital period as it spirals inwards.

The 'lunar cataclysm', 3.8 billion years ago

Returning to the Earth-Moon system, we now consider the significance of the 3.8 billion year point (c) in Fig. 2.3. This is the date of a 'lunar cataclysm' revealed by measurements of isotopes in the lunar rocks returned by the Apollo programme [2.4]. The Moon was extensively resurfaced at that time by a massive bombardment that produced the major craters (and probably many small ones). This was 600 million or 700 million years after the formation of the Earth and Moon. Planetary accretion does not take that long and the cataclysm was not a final stage in accretion of the Moon but a distinct event occurring over a limited time interval [2.5], after the Moon had been complete and quiescent for hundreds of millions of years, with an established crust, some of which survived the cataclysm. The nature of the bombardment has been subject to much debate, almost always assuming that it was not restricted to the Moon, but that it also stirred up the surface of the Earth (and even the other terrestrial planets). But the evidence contradicts this assumption and the cataclysm is better explained as a merger of two satellites of the Earth that were brought together by orbital evolution, driven by tidal friction, after 600 million years or so in independent orbits. The smaller body did not simply impact the Moon, but broke up, so that its fragments bombarded the Moon independently. The mechanism by which this occurred is discussed presently.

The date of what is referred to as the late heavy bombardment (LHB) of the Moon has been determined from isotopes in impact melts, glassy globules thrown up by impacts that were violent enough to cause some melting. The impact melts (impactites) brought back with the Apollo samples are all dated in the range 3.8 to 3.9 billion years ago (Fig. 2.5). They are also found in meteorites of lunar origin, which show a wider range of dates, but as these were thrown off the Moon by later impacts that would have caused some resetting of the isotopic clocks, we discount them in assessing the LHB date.

The Earth had a second satellite for 600 million years

Oxygen isotopes offer evidence of where the impactor(s) came from. The mass 16 isotope, ^{16}O, is the most abundant, but with about 0.2% of ^{18}O and 0.04% of ^{17}O. The different masses give them slightly different properties, so that in chemical reactions, such as the formation of minerals, as well as physical processes, such as the evaporation of water, there is a slight temperature dependent separation of the isotopes. This is termed mass-dependent fractionation, which is discussed further in Chapter 3, in connection with the temperature dependent partitioning of oxygen isotopes between sea water and carbonate precipitated on the sea

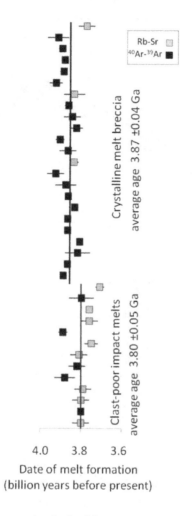

Fig. 2.5. Dates of lunar impact melts obtained from measurements of the isotopes in two different radioactive decays. From data by Papike et al [2.6]. The limited range of ages indicates that the impacts occurred within a short time interval and that they had a common cause. Two groups of impact melts are distinguished. Crystalline melt breccias are of impact origin but contain unmelted fragments (clastics) that were incorporated in the melt and may bias the age estimates towards earlier dates. The clastic poor impact melts are less abundant, but are presumed to give more reliable impact dates. The difference is slight, but on this basis we favour a date 3.8 Ga (billion years) ago rather than the more often quoted 3.9 Ga.

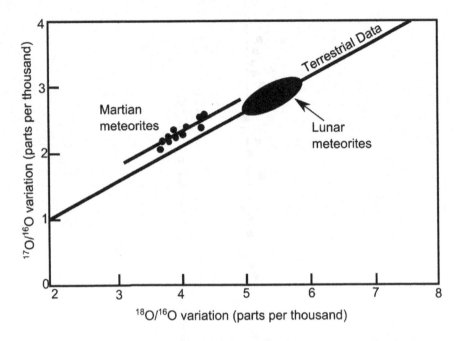

Fig. 2.6. Mass-dependent fractionation of oxygen isotopes. This shows the departures from a reference standard of ^{17}O and ^{18}O abundances for samples from the Earth and from meteorites of lunar and Martian origins. The lunar meteorites fall on the terrestrial fractionation line, obtained from a wide range of rocks on the Earth, but meteorites of Martian origin have their own parallel fractionation line, indicating slightly different isotopic compositions of the starting materials of the two planets. After McSween [1.8].

floor. Most often measured is the fractional departure from a standard reference of the proportion of ^{18}O in a sample. This is given the symbol $\delta^{18}O$ and is always reported as parts-per-thousand [2.7]. For all terrestrial samples, the variations in the corresponding quantity for ^{17}O, that is $\delta^{17}O$, are half of the variations in $\delta^{18}O$ because the mass difference from ^{16}O is half of the $^{18}O-^{16}O$ mass difference. If various mass-dependent fractionation processes cause partial separation of ^{18}O from ^{16}O, they also cause separations of ^{17}O from ^{16}O that are half as strong and a graph of $\delta^{17}O$ vs $\delta^{18}O$ is a straight line with a gradient of ½ (Fig. 2.6).

Isotopes from lunar meteorites fall on the terrestrial fractionation line but Martian meteorites depart systematically from it and have their own parallel fractionation line. Fractionation processes on Mars operated in the same way as on Earth, but they started with a slightly different isotopic mix. This is one of the indications of oxygen isotopic variations in the solar system; meteorites show similar differences. The fact that lunar samples fall on the terrestrial fractionation line indicates that the lunar material accreted at the same distance from the Sun as the Earth. On its own, this is as expected, but since the lunar surface is mixed with debris from the impactors, that must be true for the impactors as well as for the bulk of the Moon. But the impactor material could not have remained in independent solar orbit(s) close to that of the Earth for 600 million years. It would have been collected by the Earth and Moon in much less time than that. The material must have been safely stored in a terrestrial orbit, as a second satellite, sufficiently separated from the Moon to escape interaction until tidal evolution of the orbits brought them together.

The youthful Earth may have had several satellites, but we see that only two of them could have remained independent for 600 million years. Almost certainly the larger one (which we now identify as the Moon) was in a closer orbit, because the rate of orbital evolution is proportional to the satellite mass. Being then much closer to the Earth than now, its orbit would have expanded quite rapidly by tidal friction, so that it approached the orbit of the smaller one. This disturbed the orbit of the minor one, causing it to approach the Moon close enough to make it gravitationally unstable. This instability is known as the Roche limit problem, after E. Roche, who first drew attention to it in 1850. If the two interacting bodies had densities equal to that of the Moon, as was probably the case, then the self gravitation of the smaller one would have been insufficient to hold it together if it approached the larger one within about 2.4 times the radius of the large one, and it would have broken up. The probability of coming within this distance is proportional to the area of a circle of that radius and is therefore $2.4^2 = 5.8$ times more likely to occur than a direct impact. Instead of simply hitting the Moon, the smaller body, which we refer to as the impactor, broke up. Its fragments remained in terrestrial orbits close to that of the Moon, bombarding it over a period that could have been millions of years and involved secondary fragmentations. It is not completely impossible that, after sling-shot passes of the Moon, a few fragments escaped into solar orbits, from which they could have impacted the Earth, but the energy requirement would have made that unlikely. There is no reason to suppose that the Earth was subjected to a late heavy bombardment. That was restricted to the Moon, with no significant consequences elsewhere in the solar system, although confirmation by geology surviving from 3.8 billion years ago is admittedly limited.

An essential feature of this explanation of the late bombardment is that the impacts were at 'low' velocity, little more than the escape speed of the Moon, 2.4 km/s, and nothing like the speeds of impactors arriving from orbits extending to the asteroidal belt (15 to 20 km/s). This accords with an observation reported by P. H. Warren and co-authors as a paradox, that the lunar impactites had very little evidence of vaporisation, less than 0.1% of the melt volumes [2.8]. They are, in this respect, quite different from debris from higher velocity (asteroidal) impacts, which are energetic enough to produce not just melts, but vapour, apparent as condensate with volumes typically 11% of the melt volumes. The sizes of the lunar craters demand massive impacts, but not high velocity impacts. The total mass of impacting material is not well constrained, but may have been as little as 1% of the mass of the Moon. However, the postulated second satellite must have been large enough to hold itself together gravitationally, so that an approach within the Moon's Roche limiting distance caused its break-up.

Impacts on Mercury

Images of Mercury show craters similar to those seen on the Moon. If, as we suggest, Mercury once had a satellite, or satellites, that spiralled into the planet by tidal friction, they would have broken up at the Roche limiting distance and bombarded Mercury with their fragments. If it is ever possible to retrieve surface samples from Mercury, then it will be possible to check on the conjecture that its craters are also results of relatively low-velocity impacts of debris from one or more former satellites that bombarded the planet at a specific time or times.

An implication of the solar tide

The tide raised in the Earth by the Sun is smaller than that caused by the Moon, but only by a factor 0.46 at the present time. In the next chapter we consider the consequences of the solar tide if the mass of the Sun were significantly different from that of the one we have, with the conclusion that, taken together, the solar tide and solar radiation impose a tight restriction on the range of masses that the Sun may have for the Earth to be habitable. If it were significantly smaller, then to receive sufficient sunlight the Earth would need to be closer but tidal friction would have stopped it rotating, and if the Sun were much larger it would have burned out too quickly for life on Earth to have developed.

Formation of the Earth and Moon required no special physical processes

The essential conclusion from the evidence discussed in this chapter is that we see no reason to suppose that the Earth and Moon are in any way anomalous or require special explanations for their origin and history. From the tidal record, we estimate that the Moon formed (at the same time as the Earth) at a distance of 20 to 30 Earth radii, as indicated by its starting point in Fig. 2.3. It was not a much closer accumulation of debris thrown up by a giant impact on the Earth, as has been popularly supposed [2.9]. The fact that we have only one satellite is a consequence of tidal friction, as is the absence of satellites of Venus and Mercury. It is obviously important that the Earth is still rotating; as mentioned, if the Moon were significantly bigger than it is (and not much more remote) then the Earth would present a constant face to it, rotating with a month-long day. It may also be important that the Earth is no closer than it is to Jupiter. Proximity to Jupiter could be responsible for the absence of large regular satellites of Mars. While the habitability of the Earth and the development of biological life require some particular physical conditions, considered here and in the next chapter, they are all consistent with normal planetary development.

George Darwin, 1845–1912
A son of Charles Darwin, qualified as a lawyer, but turned to science as a mathematician and astronomer. His enduring legacy is the theory of tides, including tidal friction and the evolution of the Moon's orbit.

3. The variable Sun and other astronomical effects

The effects on global temperature of short-term variations in sunlight

The response of the Earth to variations in insolation (the energy of solar radiation falling on it) is quite complicated and it is helpful to have direct evidence before discussing orbital effects, changes in the solar output or greenhouse warming. Climatic effects are observed at three regular periods, the year, the month and, most surprisingly, the week. These periods are unrelated to one another, or to properties of the Sun, and have very different effects. The regular repetitions provide evidence of the surety of the observations and some understanding of them is pre-requisite to confidence in explanations of longer term effects.

The annual cycle in sunlight received depends very strongly on latitude, and the resulting seasonal effects are so strong that it is difficult to discern a global cycle of average temperature. However, the Earth's orbit is elliptical, bringing it 3.4% closer to the Sun on January 4, the height of summer in the southern hemisphere, than it is on July 4, the height of the northern hemisphere summer. Since the total insolation varies inversely as the square of distance from the Sun, 6.9% more sunlight reaches the Earth on January 4 than on July 4. The effect of this on the temperature variation of the Earth is discussed in terms of fundamental radiation laws, often referred to as black body radiation, the essential features of which are presented in Note [3.1] and discussed further in Chapter 8. In the absence of complicating influences it would cause a global mean surface temperature 4.8°C higher on January 4 [3.2],

Max Planck, 1858–1947

A German theoretical physicist whose theory of radiation (Planck's law) became the basis of quantum mechanics. He was a very conservative scientist; reluctant to accept his own revolutionary theory although recognition of its significance eventually won him a Nobel Prize. As Planck acknowledged, the idea of energy quantization in discrete units originated with Ludwig Boltzmann's study of heat. The Planck theory was completed by Einstein's evidence of the quantization of light from the photoelectric effect. The fundamental coefficient of quantum physics, Planck's constant, h, relates the energy of a quantum of radiation (photon) to its frequency.

Milutin Milankovitch (Milanković), 1879–1958

A Serbian civil engineer who progressively turned to fundamental studies in astronomy and climatology when appointed to a chair in Belgrade. The idea that ice ages were driven by changes in the Earth's orbital motion had been discussed extensively by James Croll, in England, and others for many years, but a quantitative theory awaited the mathematical insight of Milankovitch, whose name is remembered in the Milankovitch cycles of climate variation. Portrait by Paja Jovanović.

but that is not observed, for several reasons. The most important is that the hemispheres are very different, with most of the land in the north and oceans concentrated in the south, making the thermal inertia of the oceans more effective in moderating temperature variations in the south, as well as making it cloudier. But the Earth is orbiting faster when it is closer to the Sun, making the southern hemisphere summer slightly shorter and the winter longer, with the reverse occurring in the north. On a seasonal basis this compensates for the different intensities of solar radiation, because the total radiation received over the course of a year is the same over both hemispheres [3.3]. The climatic differences between them are not caused by a difference in total radiation, but by the seasonal timing of its maxima and the land/ocean distribution.

The elliptical orbit about the Sun is not strictly the orbit of the Earth but the orbit of the Earth-Moon combination. They are each in orbit about their shared centre of mass: a point inside the Earth 4671 km from its centre. It is this centre of mass that follows the elliptical orbit about the Sun. The Earth is oscillating about this solar orbital path with the 29.53-day period of the lunar month (Fig. 3.1). The peak-to-peak amplitude of the oscillation is 2×4671 km = 9342 km, which is 62 millionths of the Earth-Sun distance, very much smaller than the 5 million kilometre oscillation due to the ellipticity of the orbit about the Sun. Nevertheless, the lunar period has been seen in satellite observations of atmospheric temperature, using thermal radiation, at microwave frequencies, from oxygen in the lower atmosphere [3.4]. The scientific significance of this effect is that its period is unique. It cannot be attributed to anything but the Earth-Moon motion, and the temperature oscillation must be explained in terms of that motion even though it is observed against a background of bigger effects. Although the average magnitude for the Earth as a whole (about 0.01°) agrees reasonably with the black body model, as in Note [3.4], it is not really just a simple radiation balance. There is a complicated latitude dependence, with the expected temperature variations at high and low latitudes but opposite variations at intermediate latitudes. Also, temperature variation generally occurs ahead of the sunlight intensity (insolation), so that it is not the result of a simple radiation balance, but indicates variations in atmospheric circulation, possibly involving the atmospheric tide. However, the fact that an average effect of the magnitude expected by a simple theory is observed in a repeated cycle of insolation encourages the view that global temperature responds in a comprehensible way to variations in the input of solar radiation.

The same satellite observations have been analysed for evidence of a weekly cycle, which is apparent in the northern hemisphere but is very weak or absent in the south [3.5]. The 7-day week is not a unit of time related to any natural astronomical or terrestrial cycle; it is a cycle of human activity and its appearance in the temperature record must be interpreted as a consequence of that activity. Weekdays are seen to be

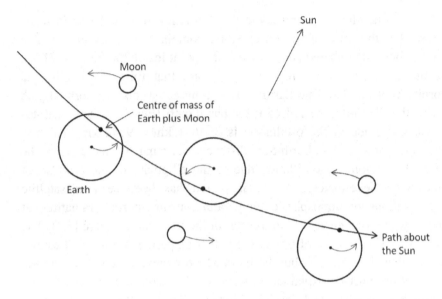

Fig. 3.1. Earth and Moon are in orbit about their combined centre of mass, a point within the Earth, and this centre of mass follows an elliptical path about the Sun. This results in a monthly oscillation in the Earth-Sun distance, with a detectable effect on atmospheric temperature.

warmer than weekends. Although details of the causal connection remain to be clarified, this observation provides incontrovertible evidence that human activity directly affects atmospheric temperature. If CO_2 is the link, then a weekly variation in its distribution, perhaps by vertical mixing, must be responsible because, on a hemispheric scale, the total atmospheric content cannot plausibly fluctuate so rapidly. The redistribution of CO_2 is nevertheless the most plausible explanation because variations in the heat release by human activity appear to be too small [3.6]. These monthly and weekly cycles are, of course, very rapid compared with the climate changes that must now be contemplated but, since they are repeated with characteristic periods, they leave no doubt about the ultimate causes of the observed effects. Although they are not simple, they allow the conclusion that longer term effects that lack the control of precise repetition or have possible multiple causes, are also comprehensible.

Orbital effects

Climatic effects occurring over tens of thousands of years are caused by cyclic variations in the Earth's orbit and the orientation of the axis of rotation that are driven primarily by gravitational interactions with the Sun and Moon, with smaller effects of interactions with the other planets. In the second century BC, Hipparchus made a remarkably accurate estimate of the rate of variation in the orientation of the rotational axis, which we now understand as the precession of the Earth, which rotates in the manner of a top. It is the most striking of the rotational and orbital variations. The Earth has an equatorial bulge, with an equatorial radius 21 km (0.003%) bigger than its polar radius. This is the result of a balance between the centrifugal effect of rotation and self-gravitation, which pulls it towards a spherical shape. The precession is a consequence of the gravitational interaction of the Sun and Moon with the bulge. The equator is inclined to the plane of the Earth's orbit by 23.45°, giving us the seasons. Each hemisphere is inclined towards the Sun for half of every year, but the gravitational interactions cause the rotational axis to precess in a conical path about the normal to the orbital plane, with a period of 25,700 years. The motion is modified by a progressive change in the orientation of the axis of the orbital ellipse, resulting in a somewhat shorter climatic precessional period of 21,000 years. This means that, whereas the Earth is now closest to the Sun in the southern hemisphere summer, in 10,500 years time it will be closest in the northern hemisphere summer (as it was 10,500 years ago). This was understood in the mid-19th century, when recognition of the occurrence of ice ages invited the inference that such climatic effects are controlled by orbital variations. The idea remained in limbo until the 1940s, when M. Milankovitch gave it a sound theoretical basis and the periodic orbital variations are known as Milankovitch cycles. There are periods of 41,000 years and 100,000 years, as well as 21,000 years, but they are not completely independent, and interfere with one another, resulting in complicated climatic variations that require numerical work to unravel.

There are several observational tests of Milankovitch periodicity in climate relying on the dating of events such as advances and retreats of glaciers. The general conclusion is that climatic records do show the Milankovitch periods, but the correlation is very imperfect and there are other important effects. Particular attention has been given to variations in the relative abundances of oxygen isotopes in the fossilised shells of marine creatures, which offer an extended record of ocean temperature. The use of oxygen isotopes as tracers of geological activity is referred to in Chapter 2. Figure 2.6 shows plots of isotopic ratios that illustrate the phenomenon of mass-dependent fractionation, caused by subtle differences between the physical and chemical properties of different isotopes, arising from their different masses. The differences depend on temperature as well as chemical environment. Shells of marine creatures have slightly higher proportions of the heavy isotope ^{18}O than the sea water in which they grow. Since there is no direct measure of the isotopic ratio of the water at the time when ancient fossil shells grew, the measurements are compared with a reference standard, known by its acronym SMOW (Standard Mean Ocean Water), and the results are represented by the symbol $\delta^{18}O$ [2.7], which is a quantitative measure of the difference in ^{18}O abundance from SMOW. $\delta^{18}O$ decreases with the temperature of the water in which shells grow and is therefore a measure of the temperature at the time. The assumption that the composition of the water coincided with SMOW is obviously not valid and there are methods of correcting for this problem when two different carbonate minerals are formed at the same time but, for a qualitative indication of temperature variations with time, we do not need a correction. Water evaporating from the oceans is relatively low in ^{18}O because the lighter molecules evaporate more readily and, when it falls as snow in polar regions, the accumulating ice is low in ^{18}O. During cold periods, when a large volume of water is locked up as ice in polar caps, it leaves the oceans with a high ^{18}O content. This causes high $\delta^{18}O$ values in shells grown at the time, reinforcing the high $\delta^{18}O$ arising from growth in cool conditions. Identification of cool and warm periods from oxygen isotopes in carbonates is therefore unambiguous, although absolute temperatures are not obtainable without additional information.

Putting together the various indicators of past climates, it is evident that the Milankovitch periods have a real effect. They have even been seen in 440 million year old sediments. But they are all tens of thousands of years long and offer no explanation of either glaciations lasting for millions of years or the fluctuations occurring over a few centuries that are apparent in historical and archaeological records. Possibly the most significant astronomical driver of century time scale climate change is the intrinsic variability of the Sun, and we refer below, and in Chapter 4, to the possibility that very prolonged glaciations are caused by the impact on the Sun's magnetic field of variations in the density of material in interstellar space.

Solar variability, sunspots and the little ice age

The Sun has an internal cycle, averaging about 11 years, which has been observed over several centuries by the appearance and disappearance of sunspots (Fig. 3.2). Over the last 30 years, satellite measurements of solar luminosity have shown that the Sun emits more radiation at times of sunspot maxima than when there are few spots. The total radiant energy varies by about 0.12%, but ultraviolet (UV) radiation varies more dramatically (as does the particle emission). Although the variation in total radiation is expected to cause an 11-year oscillation of about 0.1° in global mean temperature, the variation in UV radiation is more significant. Circulation of the stratosphere is strongly affected by it, most noticeably with major temperature variations in the polar stratosphere correlated with the sunspot cycle, an effect mentioned also in Chapter 7. Since the whole atmosphere is influenced by the stratospheric circulation, UV radiation has a much stronger effect than its total energy suggests. As in the case of the lunar cycle, a simple energy balance calculation gives a very incomplete account of climatic

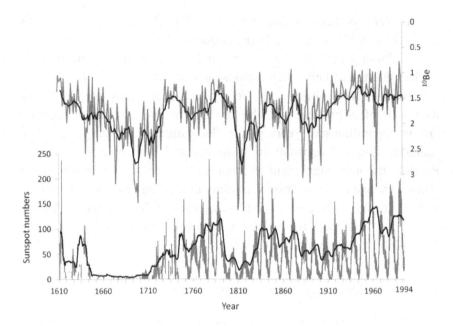

Fig. 3.2. A comparison of sunspot numbers and beryllium-10 concentrations in the NGRIP (Greenland) ice core. Light lines show annual averages and heavier lines are 11-year running means. Sunspot data from NASA and beryllium data reported by the North Greenland Ice Core Project (NGRIP), authors A.–M. Berggren et al. The significance of the beryllium data is discussed in Chapter 4.

effects and the less direct effect on atmospheric circulation is at least as important. A conclusion from this is that climate changes arising from varying insolation are very latitude-dependent, but the black body model nevertheless gives a useful measure of the global average effect. It is significant that satellite observations of solar luminosity now extend over 3 sunspot cycles and show no evidence of a trend superimposed on the 11-year cycle. A suggestion that modern global warming can be attributed to a varying solar output has no validity.

Evidence that the luminosity of the Sun is correlated with the sunspot cycle is strong enough to justify the inference of luminosity variations associated with sunspot numbers in the historical record, long before there were instrumental observations of the solar output. Of particular interest is the 70-year period (1645–1715) known as the Maunder minimum, when there were hardly any sunspots [3.7]. This is discussed further in Chapter 4. It occurred near to the end of a period of lowered global temperature, 'the little ice age' from the mid-13th century to mid-18th century. It is generally assumed that the Maunder minimum and the little ice age were causally related and, although convincing evidence is not now obtainable, the coincidence in timing combined with the modern demonstration of the variation of solar luminosity through the sunspot cycle provide strong circumstantial evidence. Was the Maunder minimum the finale to a more extended period of reduced solar activity? Interruption of the sunspot cycle may be a recurring phenomenon, but reports of earlier prolonged sunspot minima are not as well documented. There is also a suggestion that the duration of the sunspot cycle is an indicator of output variations. It is not a perfectly regular cycle; the period averages about 11 years but can be two or three years longer. These effects are internal to the Sun but solar physics is not adequate to predict them. Another suggestion that has arisen is that, as the solar system moves through a background of galactic dust and gas of variable density, some of it falls into the Sun, increasing the surface brightness. But to be effective, this would require an implausible rate of increase in the mass of the Sun [3.8] and we discount it (as did Kelvin in the mid 1800s). However, variations in the density of the galactic medium surrounding the solar system may have a more subtle, but still significant, climatic effect as mentioned in Chapter 4 in connection with variations in the terrestrial and solar magnetic fields.

The faint early sun paradox

An astronomical problem that has attracted considerable attention because of its relevance to the Earth's climate, but remains enigmatic, is referred to as the 'faint early Sun paradox'. This concerns the nuclear reactions in the Sun's core that produce its energy. The net effect on the composition is the conversion of hydrogen (^{1}H) to helium (^{4}He), with four atoms of ^{1}H consumed for each atom of ^{4}He produced. The very large energy release is illustrated by the wide separation of ^{1}H and ^{4}He in

Fig. 1.1. The reactions occur because, when energetic protons (^{1}H nuclei) collide, they may combine to form deuterium (^{2}H) and further bombardment converts that to ^{3}He. Collisions between these atoms then produce ^{4}He, with release of surplus protons. A very high temperature (of order 15 million degrees) is required because these fusion reactions only occur when the interacting nuclei collide with high energies and the rate of reaction increases with temperature. Other nuclear cycles become important at higher temperatures, but the process is self-stabilising and does not generate indefinitely increasing heat because that causes higher temperatures and consequent thermal expansion, reducing the concentration of interacting atoms. But the helium is intrinsically denser than the hydrogen that it replaces, so, as its abundance increases, the Sun's core contracts, causing an increase in pressure, with a consequent rise in temperature and accelerated nuclear reactions. Solar model calculations, using the detailed nuclear data that are available, indicate that 4 billion years ago the Sun would have been 30% less luminous than it is now. This is the faint early Sun model. The paradox is that, with solar radiation so much weaker, it is presumed that the Earth would have been frozen hard, with a mean surface temperature at least 23°C lower than at present [10.5], whereas geological evidence points to a liquid ocean at that time.

The standard model of the Sun, in its present state, is quite well constrained by two kinds of observation. By analogy with the seismological study of the Earth's structure, which uses waves and vibrations generated by earthquakes, helioseismology uses vibrations of the Sun, apparent as fluctuations of the light output, to model the internal structure. The energies of neutrinos of solar origin identify the nuclear reactions and so give a measure of the deep internal temperature. The direct observational evidence of both structure and temperature gives confidence in the solar model. Moreover, we can note that details of the solar and nuclear models are not critical; it is the changes to them that interest us, and the only changes are consequences of the exchange of the denser helium for some of the hydrogen. The paradox arises because this exchange causes an increase in the density of the Sun's core. It would be resolved if the total mass of the Sun decreased with time, reducing the central compression. We consider that possibility here but discuss alternatives in Chapter 10. The youthful Sun would have lost material by mass ejections from its outer atmosphere, as young stars are observed to do now, but that process would have lasted no more than 100 million years and cannot explain a variation in the Sun's mass extending over 4 billion years. The continuing loss of mass by particle emission (the solar wind) accounts only for a loss of about 0.1% of the solar mass in this time. By the equivalence of mass and energy ($E = mc^2$), the Sun also loses mass at a steady rate by radiation; 0.03% of the Sun's mass has been lost in this way in 4.5 billion years. To this must be added the loss due to neutrinos that are products of the nuclear reactions and escape from the Sun. But neutrino energy is only a modest fraction of the total nuclear energy produced and we can calculate the upper limit of the total of these effects by postulating a star composed initially of 100% hydrogen that is converted to 100% helium [3.9]. In spite of exaggerating the Sun's mass loss, this is inadequate to resolve the paradox. A change exceeding 1% in the solar energy received at the Earth in response to mass loss by the Sun could not have resulted from either solar luminosity or orbit dilation. After considering the alternatives in Chapter 10, we conclude that a strong early greenhouse effect must be the favoured explanation.

The size of the Sun is critical

On the subject of tidal friction and the history of the Earth-Moon orbits (Chapter 2), solar tidal friction would take on a different role if the Sun were less massive than it is, even if only moderately so. For stars like the Sun, radiant outputs depend very strongly on their masses. Luminosity is doubled for a 22% greater mass or increased by a factor 11.3 for a doubled mass. Thus, if the Sun were smaller than it is, to be habitable the Earth would need to be closer to receive the same sunlight. But friction of the solar tide would be strongly increased at the smaller distance, varying as the sixth power (R^{-6}) of distance R, but only as the square (M^2) of the Sun's mass, M. With this very strong variation, a

Gustav Spörer, 1822–1895

A German astronomer who spent many years studying the Sun, sunspots and records of observations predating his own, especially those by the English astronomer, Richard Carrington. He appears to have been the first to notice an extended period of reduced solar activity, with very few sunspots from approximately 1645–1715. His report was not taken seriously at the time and that period was subsequently named the Maunder minimum. However, the validity of his conclusions was later acknowledged by naming an earlier period, 1460–1550, the Spörer minimum [3.7]. There are no sunspot records from that period but reduced solar activity was inferred by J. A. Eddy from the low abundance of the carbon isotope ^{14}C in tree rings from that time. This isotope is produced by cosmic ray bombardment of the upper atmosphere and varies with solar activity.

modest decrease in M would put the Earth into the tidal situation of Venus and Mercury, with rotation, relative to the Sun stopped. Note [3.10] concludes that this would have happened with a 20% to 25% smaller Sun, making the Earth uninhabitable at any distance from it. There would be two obvious consequences. The side facing the Sun would be a baked desert, with all of the water condensed on the deep frozen dark side. It is not obvious that, even with atmospheric circulation, there would be a usable twilight ring, because the area of total darkness would too completely freeze out all the water. Another consequence is that, with a very slowly rotating Earth, lunar tidal friction would have caused the Moon to plunge into it with the calamitous consequences that this implies.

At the other extreme, with greater radiation from a more massive Sun, its useful lifetime would be shorter because its nuclear resource would be consumed faster. Note [3.11] concludes that the Sun would already be in its death throes if it were 32% more massive. Thus, with our human-centric view of what constitutes habitable planets, one of the necessary conditions is that they orbit stars with a very limited range of masses. Although there are many stars with masses comparable to that of the Sun, and this may not appear to be a very restrictive condition, it means that, in the search for stars that may have habitable planets, the great majority are out of contention simply because of the limited range of suitable stellar sizes.

Edward Maunder, 1851–1928

English astronomer, much of whose professional life was concerned with sunspots. Building on the work of Carrington and Spörer, and working closely with his second wife, Annie, Maunder drew attention to the repeated latitude migrations of sunspots through the solar cycle, starting at mid solar latitudes (± 30°) and moving to low latitudes before fading out and starting again, 11 years later, at mid-latitudes. Maunder's documentation of the extended period of solar inactivity, 1645–1715, was, as for Spörer's work, disregarded at the time, but later confirmed as robust by J. A. Eddy, whose 1976 paper [3.7] convinced the scientific community of the significance of what he called the Maunder minimum. Interest in it arises particularly from an apparent causal connection to the 'little ice age'.

4. The magnetic field

Magnetic fields of planets

We remark in the preface that one of the features of the Earth that controls the environment is the magnetic field. A comparison with the fields of the other planets (Table 4.1) helps to put this into perspective. The strength of the Earth's field is comparable to the fields of the giant, outer planets, but is unique in the inner solar system. It is not immediately obvious why this is so. Generation of a planetary field requires large scale, complicated motion in a fluid electrical conductor. All of the terrestrial planets (Mercury, Venus, Earth, Mars), as well as the Moon, have iron cores (Fig. 4.1) that are at least partly liquid and the pressures in the interiors of the gas-rich giants suffice to convert materials such as hydrogen to metallic forms, so the fluid electrical conductor requirement is satisfied in all cases. The more difficult parts of the problem are to determine the nature of the fluid motion that drives spontaneous dynamo action and, most importantly, the source of energy that maintains it. Although basic principles are understood, many details remain uncertain. Nevertheless we need to examine them to get an idea of why the Earth is the only solid planet with a strong field.

Magnetic rocks

Consider first Mars and the Moon, both of which have magnetized rocks in their crusts. The rocks contain iron-bearing minerals, or, in the case of

Table 4.1. Magnetic fields of planets.

Planet	Mean surface field strength (nT)	Cause
Mercury	475	weak dynamo
Venus	none detected	probably none at all
Earth	41455	dynamo
Moon	~1	magnetized rock
Mars	3.5	magnetized rock
Jupiter	650 000	dynamo
Saturn	32850	dynamo
Uranus	32170	dynamo
Neptune	18560	dynamo

lunar rocks, very fine grains of iron or iron-nickel alloy, that give them magnetic properties similar to, although much weaker than, laboratory or industrial magnets, which remain magnetized after removal from a magnetic field. The observed external fields of Mars and the Moon are caused by these rocks and are very irregular, with patterns reflecting local geology, quite different from the global pattern of the Earth's field. Nevertheless, the fact that the rocks are magnetized means that they were once exposed to fields. Measurements on lunar samples indicate that, in at least some cases, they were magnetized by cooling in a magnetic field and, as they have not been heated since being brought back by the Apollo astronauts, it was evidently a lunar field. From less direct evidence, the same is inferred for the Martian crustal rocks. Also, it would be difficult to explain the crustal magnetizations of both Mars and the Moon unless they once had magnetic fields, although probably only transiently in the case of the Moon. However, the driving mechanisms have now failed. On the other hand, Mercury still has a weak internally driven field. Perhaps importantly, although it is a smaller planet, its core is larger than the core of Mars (Fig. 4.1). But the core of Venus is larger still, not much smaller than the Earth's core, and the apparently complete absence of a field is more difficult to explain. If Venus did once have a field, then there is no evidence now because the surface is too hot to have retained remanent magnetism, as have surface rocks on Mars and the Moon.

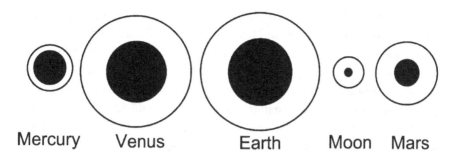

Fig. 4.1. Relative sizes of the inner four planets and the Moon, and of their iron cores (shown in black). Fluid electrically-conducting cores are necessary for magnetic fields, but Venus, Mars and the Moon have no internally-generated fields and Mercury has a very weak one. The Earth's strong field is unique in the inner solar system (Table 4.1).

Planetary dynamos

Electric currents in planetary cores must be driven by sources of energy. These are generally understood to be heat or chemical separation processes that cause convective stirring. In a convectively stirred medium there is a necessary minimum increase in temperature with depth termed the adiabatic gradient, as discussed in Chapter 8 in connection with the temperature of the Venusian atmosphere. Magnetic dynamos in planetary cores require 3-dimensional stirring of the cores, which must therefore support temperature gradients that are close to adiabatic. This means that they lose heat at a rate that at least suffices to maintain that temperature gradient. The heat lost by the Earth's core produces a layer of hot, buoyant material at the base of the mantle, and this rises to the surface as narrow, convective plumes, which have other significant effects (discussed in Chapter 9 and illustrated in Fig. 9.2). For this to occur, the core must be about 1000° hotter than the mantle above it. But this would not have been the case when planets first formed, so the essential requirement is that the mantle must have cooled more than the core. Thus we see another requirement for a dynamo in a planet such as Earth or Venus: a sufficiently rapidly cooling mantle. With this condition we can see a possible reason why Venus has no field. We point out, near to the end of Chapter 7, that the low abundance of argon, specifically the isotope ^{40}Ar, in the Venusian atmosphere indicates much less effective mantle convection than in the Earth, from which the escape of argon to the atmosphere is a consequence of convection. We infer that, in Venus, convective cooling of the mantle has been insufficient to develop a temperature difference from the core big enough to allow the core cooling needed for dynamo action. What could be the reason for that? Venus lacks an ocean and cannot have subducted sea water to lubricate convective motion, as in the Earth. Thus, we see another important environmental effect of the oceans: water-lubricated mantle cooling that is fast enough to allow the escape of core heat at a rate sufficient to maintain the magnetic field.

At this point we need to turn to the very interesting situation of Mercury, which is at least as dry as Venus, but has a weak dynamo. With its large core, the mantle is thin, little more than 500 km, compared with 2890 km for both the Earth and Venus. It allows more effective conduction of core heat to the surface without convection, evidently just sufficient for a thermally powered dynamo [4.1]. However, it also appears possible that the Mercury dynamo is driven by the very strong tidal forces arising from its closeness to the Sun and very elliptical orbit. Although convection in the Venus mantle must once have been effective in cooling the core, it appears not to be so now, but convection is evidently not necessary for the Mercury dynamo. The conductive regime in Mercury's mantle must have prevailed for billions of years to leave the ancient cratered surface apparently unmodified by convectively driven tectonics.

The energy source for the Earth's dynamo

The viability of the Mercurian dynamo may be a close call, but that can hardly be true of the terrestrial dynamo. The Earth has had a magnetic field comparable in strength to the present field for as long as rock magnetic measurements can discern — at least 3.5 billion years of its 4.5 billion-year existence. Most of the energy driving it is believed to be derived in two ways from the growth of the solid inner core as the core cools. The solidifying liquid releases latent heat, which is convectively transported upwards in the outer core and also the solid rejects some of the light solute that is dissolved in the outer core. This produces buoyant material at the boundary and the gravitational energy of its mixing into the outer core drives what is termed compositional convection. The amount of light material released in this way is estimated from the seismologically determined density difference between the inner and outer cores, which exceeds by a factor of 4 the density difference attributed to solidification of material of constant composition. (The density profile of the Earth is shown by the figure inside the front cover).

Compositional convection provides the dominant dynamo energy source, at least at the present time. Heat from radioactivity in the core is also commonly favoured, but with no general agreement on how important it is. Uranium and thorium are not candidates and attention has concentrated on the solubility of potassium, with its radioactive isotope ^{40}K, in liquid iron alloy at high pressure. Most measurements agree that there is almost certainly some potassium in the core but that the quantity is small.

Another possible dynamo energy source that has strong advocates, but has also attracted serious doubt, is precession. This is the variation in orientation of the rotational axis caused by the gravitational forces of the Moon and Sun on the equatorial bulge, as discussed in Chapter 3. The axis of the Earth's field is misaligned with the geographic (rotational) axis by about 10°, but it is continually changing and there are irregular features of the field that grow, decay and drift. But, averaged over 10,000 years or more, the magnetic and geographic axes coincide. This is well documented by measurements of magnetizations of rocks and is illustrated in Fig. 4.2. Clearly the rotation of the Earth has a strong control on the geomagnetic dynamo, inviting the supposition that it does not merely bias the turbulent convective motion that generates the field but provides a driving force by a mechanism outlined in Note [4.3].

The Earth's field varies but it is robust

Whatever the details of the driving mechanism, the behaviour of the Earth's field is well documented and there is nothing in the record to indicate that it will not continue indefinitely with strength comparable to that at present. It is not steady but continually fluctuates and, on irregular occasions, reverses. In the last several million years reversals have occurred, on average, about once every 400,000 years, taking about 5000 years to occur and being very weak during a reversal. It has often been noted that the strength of the field is decreasing at the present time, but that does not mean that it is heading for a reversal. In fact, the field is slightly stronger now than the long-term average and the present rate of change can be recognised as part of the regular fluctuation, referred to as secular variation. We have no way of predicting when the next reversal will occur, although we can confidently assert that it will do so some time, perhaps in several hundred thousand years time. However, that will just be a feature of the erratic behaviour of the field when viewed on a long time scale and we must expect it to continue in this manner, probably for as long as the Earth exists. That means that heat from the core will continue to drive the convective mantle plumes that have had dramatic environmental effects at intervals of a few tens of millions of years, as discussed in Chapter 9.

The magnetosphere: extension of the Earth's field into space

The pattern of the Earth's field resembles that of a strong bar magnet or current loop at or near to the centre. Such a field is referred to as a dipole field (a magnet has two poles). This represents 80% of the average field strength at the surface, with finer features, termed the non-dipole field, accounting for about 20%. The strength of a dipole field diminishes as the cube of the distance from its source ($1/r^3$) and the strength of the non-dipole field components diminishes more strongly, so that in space the field is well represented by a dipole. It rotates with the Earth and, if the surrounding space were a perfect vacuum, the field would extend indefinitely in all directions, sweeping through space at a speed increasing with distance. But interplanetary space is occupied by a very tenuous plasma of ionized gas, which is an electrical conductor. Motion of a magnetic field relative to a conductor generates electric currents that cause forces between them with two competing effects. They tend to make the conductor move with the field and they oppose the penetration of the conductor by the field. In regions relatively near to the Earth (within about 10 times the Earth's radius), the field is strong enough to make the plasma rotate with the Earth, but at greater distances the plasma prevents the penetration by the field. The field is thus confined to a limited volume, *the magnetosphere*, within which the plasma is controlled by the field and moves with the Earth, while the plasma outside is independent of it. The magnetosphere is like a giant magnetic balloon, carrying the outermost part of the atmosphere with the Earth, as it orbits the Sun, and protecting it from the flow of the interplanetary medium.

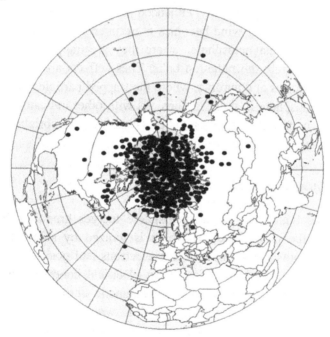

Fig. 4.2. Positions of the magnetic pole deduced from the directions of magnetic remanence in rocks up to 20 million years old. Data from Tarling [4.2].

This would be a simple picture if the interplanetary plasma were stationary, but it is not; it is flowing outwards from the Sun. This is the solar wind. It compresses the magnetosphere on the sunward side and draws it out into an extended tail on the down-wind side, resulting in the structure illustrated in Fig. 4.3. The solar wind is a highly supersonic mixture of ions, mostly protons (hydrogen nuclei) and electrons, so that a shock wave develops in it where it meets the magnetosphere. Being composed of charged particles, the solar wind is deflected by the magnetic field. It forms a turbulent layer of solar wind plasma, the magnetosheath, around the magnetosphere, at a distance of approximately 10 times the Earth's radius. Although it prevents direct interaction of the atmosphere with the solar wind, the protection is not complete. Some solar particles enter the magnetosphere via its tail and at the cusps in its boundary that mark the positions of the magnetic poles. These effects are dramatically enhanced at times of increased solar activity, when the solar wind becomes more intense, injecting energetic particles into the magnetosphere, where they are trapped in spiralling orbits that constitute the radiation belts. These effects are accompanied by magnetic storms and aurorae, where the trapped particles impinge on the upper atmosphere in polar regions. As with other indicators of solar activity, such as the sunspot number (Fig. 3.2), the brightness and latitude range of aurorae vary with an 11-year cycle.

The ionosphere and the ozone layer

At its lower boundary the magnetosphere merges gradually into the atmosphere, the upper layer of which is ionized by solar ultraviolet light, which splits the atoms into free electrons and positively charged ions. This layer, the ionosphere, is an electrical conductor in which electric currents are concentrated during magnetic storms. The ionization is negligible below about 60 km elevation, with a series of layers of increasing ionization but decreasing density up to about 300 km. Above about 500 km there is only the very tenuous, but fully ionized magnetospheric plasma, composed mostly of hydrogen (as separate protons and electrons). The magnetosphere protects the atmosphere not just from the solar wind but from more energetic charged particles approaching the Earth, by deflecting all but the most energetic cosmic rays. Like the solar wind, cosmic rays are charged particles, mostly protons, but they have much higher energies and are less effectively deflected by the magnetic field. Some originate from the Sun, but the more energetic ones have sources in remote parts of the galaxy,

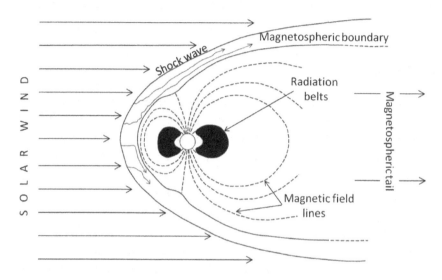

Fig. 4.3. A diagrammatic cross section of the magnetosphere.

or beyond. However, the magnetic field does not affect radiation (optical, infrared, ultraviolet or X-rays) and the magnetosphere is transparent to sunlight and to solar ultraviolet radiation. This reaches the ionosphere, where much of it is absorbed, causing ionization. The Earth is protected from an excess of ultraviolet radiation by the upper atmosphere, including the ionosphere, and not by the magnetosphere. There are several absorption processes by which various atmospheric constituents remove ultraviolet radiation of different wavelengths. One which has attracted particular attention involves oxygen, some of which is dissociated and produces free oxygen ions that become attached to normal oxygen molecules (O_2), producing ozone (O_3). Ozone is an important ultraviolet absorber, but is readily destroyed by traces of halogen gases (fluorine, chlorine), which are released by photochemical breakdown in the stratosphere of long-lived common refrigerants, mostly chlorofluorocarbons — use of which has been reduced by international agreement. Ozone is most concentrated at an altitude of about 25 km, well below the ionosphere. It is a greenhouse gas but, like CO_2, at the low density of the stratosphere it contributes to a negative greenhouse effect, causing cooling rather than warming. This is discussed in Chapter 8 (see especially Fig. 8.4) and we refer to ozone also in Chapter 7.

The heliosphere

Although the role of the magnetosphere in screening the Earth from the solar wind and low to moderate energy cosmic rays has been recognised for many years, there is another layer of protection — the significance of which has not been as widely recognised. The Sun's magnetic field extends out past all the planets, enclosing a 'balloon' of plasma (termed the heliosphere) in the same way as the Earth's field isolates the magnetosphere. It holds the interstellar medium at a distance and protects the solar system from all but the most energetic galactic and inter-galactic cosmic rays. During solar quiet times, there is a more or less steady solar wind and magnetic field, decreasing in strength with distance from the Sun and exerting a pressure on the heliospheric boundary that balances the pressure of the interstellar medium. In disturbed times the Sun throws off vortices of higher energy plasma with their magnetic fields, compressing the magnetosphere and dilating the heliosphere. The vortices reach the Earth in a few days, but take months to reach the heliospheric boundary, which is 50 or 60 times as far away from the Sun. Thus, solar disturbances have two opposite effects in protecting the Earth and atmosphere from cosmic ray bombardment. The first to occur is a decrease in protection by compression of the magnetosphere, reducing the extent of the Earth's magnetic field, and that is followed, with a delay, by the increased protection of the dilated heliosphere. On a time scale of hours to days, variations in the intensity of the cosmic ray flux correlate with magnetic storm effects that mark magnetospheric disturbances, but much longer term changes arising from heliospheric disturbances may be more significant.

The few decades of direct instrumental recording of cosmic rays do not give information on the extended time scale that we need to assess the heliospheric control. We appeal to indirect evidence, using data on a cosmogenic isotope, ^{10}Be, produced by cosmic ray bombardment of the upper atmosphere. This is a radioactive isotope of beryllium, with a half-life of 1.39 million years, which has several geological and environmental applications. Beryllium is a metal and atoms of it produced in the upper atmosphere rapidly attach to any available dust particles and are brought down to the Earth in rain or snow. In Chapter 5 we discuss the evidence of plate tectonics and volcanism revealed by the accumulation of ^{10}Be in deep sea sediment. It is also deposited in Greenland and Antarctic ice, which occurs in datable layers, so that measured variations in the ^{10}Be concentration indicate timed variations in the intensity of the cosmic ray bombardment of the atmosphere. The precipitation is not an instantaneous process, and cannot give information about day-to-day variations related to the magnetosphere, but it is rapid enough to indicate the 11-year solar cycle (Fig. 3.2). The ^{10}Be concentration and the sunspot number at the time of its deposition are negatively correlated. This is opposite to what would happen with magnetospheric control of the cosmic ray flux, because the compressed magnetosphere offers reduced protection from cosmic rays at times of strong solar activity. Heliospheric control is indicated [4.4]. This is not unexpected because production of ^{10}Be requires cosmic rays energetic enough to cause spallation of atmospheric atoms (nitrogen, oxygen), breaking off fragments (protons, neutrons — even helium nuclei). These very high energy cosmic rays have remote origins.

Effects of heliospheric compression

With the evidence of the importance of the heliosphere to the Earth environment, we can consider further effects that it might explain. The 2008–2010 minimum in solar activity was both deeper and more prolonged than other recent minima, so we can look for effects that have become more noticeable at this time. One such is an increase in the appearance of noctilucent clouds. These are clouds of very fine ice crystals that appear in summer at high latitudes, at altitudes of about 80 km, far above normal clouds and high enough to be illuminated by sunlight when the Sun is well below the horizon. Their significance is discussed in Chapter 7. The processes by which noctilucent clouds form are not well understood, so the inference of a direct causal connection between their increased occurrence and the weakened heliospheric field, with an enhanced flux of galactic cosmic rays, remains conjectural. At least as likely is a general increase in the clouds and their extension to lower latitudes, as a consequence of increased atmospheric water vapour driven by global warming, which is most obvious at the high northern latitudes where noctilucent clouds are most often seen. However, evidence of the influence of galactic cosmic rays on cloud cover at much lower elevations, not restricted to high latitudes, has been reported [4.5] and would have a climatically significant cooling effect. Although serious cooling by noctilucent clouds would not be expected, at their high altitudes the production of water condensation nuclei by energetic cosmic rays is possible. The occurrence of climatic effects driven by solar activity, is attested by the global cooling accompanying the Maunder minimum in sunspot activity discussed in Chapter 3. We accept that as an empirical observation, which is well documented, and consider the possibility that it was, at least partly, a consequence of a contraction of the heliosphere with a weakened solar field. If so, there are implications of further consequences.

The extent of the heliosphere is controlled by a balance at its outer boundary between the magnetically controlled pressures of the solar wind and the interstellar plasma. The variations that are indicated by the ^{10}Be data are attributed to the cycle of solar activity with contraction of the heliosphere during magnetically quiet times on the Sun. In the longer term, changes in the interstellar medium, between the Sun and other stars in the galaxy, must also be contemplated. It is not a homogeneous and passive 'atmosphere' immersing the solar system, but one that has heterogeneities in which motion is controlled by magnetic fields in the same way as the solar wind. The scale is very much greater and the variations that occur in the immediate vicinity of the solar system are correspondingly slower. Prolonged increases in the local density of the interstellar medium would compress the heliosphere, in a manner similar to, but plausibly much stronger and certainly much longer-lived, than contraction of the heliosphere during periods of low solar activity. If the global cooling accompanying the Maunder minimum was due as much to contraction of the heliosphere, with a weakened solar field, as to reduced luminosity of the Sun, it would be a consequence of an increase in cloudiness resulting from increased exposure of the atmosphere to high energy cosmic rays. With this interpretation, a strong interstellar compression of the heliospheric boundary could lead to global glaciation on a multi-million year time scale. This is a plausible explanation for the extended glacial epochs that have occurred, but are otherwise unexplained. However, there are no observations to confirm the mechanism.

Walter Elsasser, 1904–1991

A theoretical physicist who had a distinguished career in atomic and quantum physics in Europe before migrating to the USA where his attention turned to geophysical problems and later to theoretical biology. He is best known for his work on magnetohydrodynamics, which he used to pioneer the theory of a self-exciting dynamo, as the origin of the Earth's magnetic field. All subsequent work has followed the principles which he established.

Edward Bullard (Sir Edward), 1907–1980

Originally trained as an atomic physicist, he turned to geophysics and made contributions over a wide range, stimulating the work of many others, especially in marine geophysics and geomagnetism. He was quick to recognise the significance of Elsasser's ideas on geomagnetic dynamo theory and produced the first mathematical model of the geodynamo. Although best known for this, his work on heat flow, particularly through the ocean floor, has been at least as important to development of geophysics.

The possibility of direct biological (evolutionary) effects of cosmic rays has often been considered, but is generally regarded as insignificant. The idea first arose, not with heliospheric considerations, but with reversals of the Earth's magnetic field, during which the field strength is weak for a few thousand years, allowing an increase in cosmic ray flux to the atmosphere. As the study of geomagnetic reversals developed, it became clear that they have been far more numerous than was initially realised and any connection with genetic mutations, evolutionary developments or mass extinctions of species is not now seriously considered.

THE EVOLVING EARTH

5. Internal heat and the evolution of the Earth

Two independent heat sources maintain the environment

We live on and in a tiny fraction of the Earth, an accessible veneer comprising the uppermost crust, the oceans and the lower atmosphere, but they are maintained by two energy sources that are remote from us, solar energy and internal energy. Fortunately, we can only manipulate our life-supporting veneer and are unable to do anything about either of these energy sources. The reliability of the solar output and the manner of its interaction with the Earth are subjects of Chapters 3 and 4. Here we examine the Earth's internal heat and the surface processes that depend on it. But first we note that, by comparison with solar energy, the effect of internal heat on the surface temperature is completely insignificant. This is emphasised by the entries in Table 5.1. Although the original gravitational energy of accretion dominates the internal energy, it is dwarfed by the solar energy received at the surface and re-radiated to space over the lifetime of the Earth. The importance of internal heat is not its appearance as heat at the surface, but as the driver of geological processes that exert a long term control on the environment.

The role of radioactivity

Chapter 1 gives some attention to radioactivity as an internal heat source but, as we now understand and Table 5.1 emphasises, it is not the reason why the Earth's interior is hot. The heat remaining from the original energy of accretion still exceeds the energy released by radioactivity in the entire life of the Earth. We note also that, as demonstrated by the numbers in Table 5.2, the progressive decay of the radioactive elements is very slow and, even with radioactivity now dominating the heat flow from the interior, its decrease will not lead to a geologically inactive (dead) planet before the Sun gives out on us (in 4 or 5 billion years time). The interior of the Earth is hot because it has always been so. The gravitational energy of its accretion far exceeded the heat that it could retain and most of the energy was radiated away during the accretion process, leaving the present internal heat as a small residuum. Radioactivity has only been topping up the internal heat, although the topping up is becoming increasingly important as the Earth cools. It slows the mantle cooling by providing more than half of the continuing heat loss from the surface (by our estimate 61.5% at the present time).

Table 5.1. A comparison of global energies. All values in units of 10^{30} joules.

Gravitational energy released as heat by formation of the Earth	233
Present residual stored heat	13.3
Radiogenic heat in the life of the Earth	7.6
Radiogenic heat still to be released	10.9
Rotational energy	0.2
Solar energy received and re-radiated in the life of the Earth	24 000

The entry in Table 5.1 for the radiogenic heat still to be released makes the point that radioactivity is effectively with us forever, with only moderate further decline. This is because, the shorter lived of the thermally important isotopes, ^{235}U and ^{40}K, are now only minor contributors and the heat still to be released depends primarily on ^{238}U (half-life 4.5 billion years) and ^{232}Th (half-life 14 billion years), which already dominate the present radiogenic heat (Table 5.2). Especially significant in the very long term is thorium, which now produces 47% of the total radiogenic heat and 80% of what it contributed 4.5 billion years ago. Another perspective on this situation is a comparison of the present radiogenic heat, 28.2 terawatts, with the corresponding value in a billion year's time, 24.3 terawatts.

Table 5.2. Heat generation by the thermally important radioactive isotopes. This is a major component of the internal heat that drives geological processes and contributes more than half of the heat flux from the interior, but its effect on surface temperature is negligible. Solar heat at the surface is 98,000 terawatts.

Isotope	% of element	Half life (billion years)	Mass in Earth (10^{16} kg)	Heat (terawatts) 4.5×10^9 years ago	Now	In 10^9 years time
^{238}U	99.274	4.468	12.86	24.5	12.21	10.45
^{235}U	0.720	0.7038	0.094	44.4	0.53	0.20
^{232}Th	100	14.01	47.9	15.9	12.72	12.11
^{40}K	0.0117	1.25	9.00	33.0	2.72	1.56
			Total	117.8	28.2	24.3

Tectonic processes depend on sea water

The significance of internal heat is that it is conveyed to the surface by convection of the rocky mantle, with consequences at the surface that are termed plate tectonics (Fig. 5.1). The cooled surface layer of the Earth (the lithosphere) is divided into quasi-rigid plates that are in relative motion. Continents are buoyant and stay at the surface but ocean floor plates survive, on average, for only 90 million years. The ocean floors are produced as hot rock at the mid-ocean ridges and spread out from there, losing heat to the ocean as they move towards subduction zones, where they plunge back into the mantle as cooled slabs (as illustrated, cartoon style, in Fig. 9.2). Most of the heat loss by the Earth occurs in this way and the increased density caused by thermal contraction of the cooled plates gives them the negative buoyancy that forces them down and provides the driving force of mantle convection. This is convection of solid material, which is softened by its high temperature and behaves as a very viscous liquid, except for the cooled plates, which move about (and subduct) as almost rigid slabs. This style of convection is made

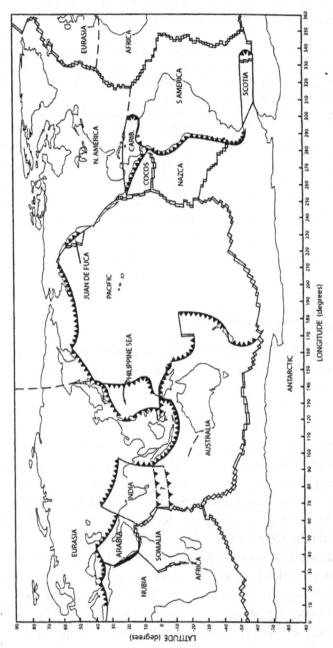

Fig. 5.1. (a) The pattern of the Earth's surface plates that are in relative motion at speeds of a few centimetres per year. 'Shark's teeth' indicate the motions of subducting plates at these boundaries.

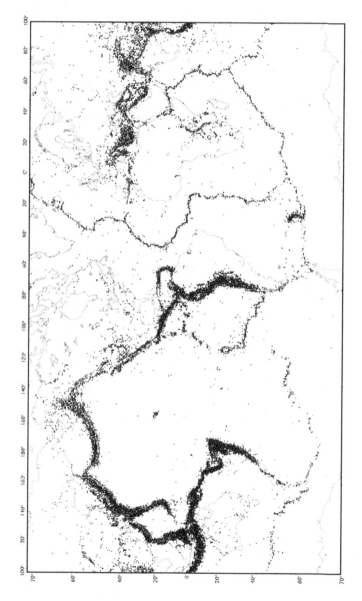

Fig.5.1.(b) Earthquakes are concentrated at the boundaries between the plates, especially convergent boundaries, where plates over-ride one another. Map shows locations of all earthquakes with magnitudes between 4 and 7 recorded between 1960 and 1990, plus those greater than 7 from 1897 (United States Geological Survey).

possible by the oceans and, within the solar system, is unique to the Earth. The minerals of the igneous ocean floor are hydrated by the sea water, which, along with wet sediment, is carried down in the subduction zones. It lowers the melting point and acts as a flux for the generation of acid (silica-rich) lava that becomes the basis of the continents. The total mass of water in the mantle may exceed that in the oceans and has a controlling influence on its mechanical properties.

In Chapter 4 we refer to ^{10}Be, a radioactive isotope of beryllium that is produced in the upper atmosphere by cosmic ray bombardment. It is quite rapidly deposited on the sea floor (and elsewhere, including polar ice) and its incorporation in deep ocean sediment provides a tracer of the recycling of sediment by mantle convection. It is found in the igneous rocks of subduction zone volcanoes, specifically fresh andesite, a rock identified with the Andes mountains. Its half-life is 1.39 million years, so it survives in the Earth only for a few million years. It presents a clear demonstration that andesite is — or incorporates — recycled ocean crust which has not spent very long buried in the mantle. The time between subduction and volcanic re-emergence is the time taken by subducting crust to reach a depth of about 100 km, where it is heated sufficiently to produce the andesitic lava. The involvement of sea water in this process is confirmed by the incorporation of the element boron in andesite. This is one of the solutes in sea water but is not a common constituent of rocks. Table 1.2 gives a representative composition of andesite, from which it is seen that the volcanic recycling of ocean floor basalt, with sea water and sediment, produces a more silica-rich composition. Also listed is a granitic composition, which is typical of the upper continental crust and is even more silica-rich. It has a more complicated origin, which is a subject of continuing debate.

The mutual control of convection and temperature

The continents and ocean basins are not fixed and immutable features of the Earth, but are continuously maintained by convective cycling of the mantle. This requires the mantle viscosity to be low enough to allow the rocky material to deform in response to convective forces and, since viscosity depends very strongly on temperature, the whole process depends on the mantle remaining hot. The heat capacity of the Earth is very large and the rate at which heat is being lost is only lowering the temperature of the mantle by about 70°C per billion years. As has been recognised for at least 200 years, the deep interior temperature is some thousands of degrees, so the cooling is not changing anything very quickly. With the continued availability of ocean water to lubricate tectonic activity, the combination of radiogenic and residual heat will suffice to ensure that geological activity will only very gradually slow down until the profligate expansion of the dying Sun engulfs the Earth (when nothing else will matter!).

When we say that almost all of the Earth's interior is hot, we need to be more specific and say how hot. For the purpose of the present discussion, 'hot' means at an absolute temperature higher than half of the local melting temperature, $T > 0.5T_M$ (absolute temperature, Kelvin, centigrade plus 273°) [5.1]. The reason for this distinction is that, at lower temperatures, rock does not gradually deform under stress, it breaks. Earthquakes occur only in rock which, by this definition, we can describe as cool. They are confined to a thin, shallow layer (the lithosphere) and the subduction zones, where slabs of lithospheric material plunge deep into the mantle, retaining their coolness to depths as great as 700 km (Fig. 5.1). The rest of the mantle is hotter and convection in it occurs by gradual deformation, without sudden breaks. No earthquakes have been detected in the lower mantle, below about 700 km depth, or in the upper mantle below the lithosphere and its extension into the subduction zones. Convection is not confined to the limited volume of the seismic zones but extends throughout the mantle, where it is controlled by the properties of hot rock. This is somewhat easier to understand than the less predictable behaviour of cool rock. The essential feature is a strong dependence of viscosity on temperature [5.2].

How does temperature dependent rock deformation control convection? And can cooling stop it? A simple argument shows that temperature and convective speed exert a mutual control on one another that stabilises both. It is easiest to understand this by including radiogenic heat in the argument, although that is not necessary to the stabilising mechanism. If convective speed becomes too slow, cooling is reduced, allowing the mantle to warm up and soften, increasing the convective speed. Conversely, convection faster than the equilibrium rate cools the mantle and slows convection. The result is a stable, equilibrium rate. But it is not a static equilibrium because heat generation decreases with time and the convective heat transfer must also do so, with convection slowing as the Earth cools. There is no end point to this process. As long as heat is available and the mantle properties do not change (which means that ocean water remains available), convection will continue at a correspondingly slowing equilibrium rate. The surface plates are moving at various speeds up to 8 cm/year, with an average of 2.6 cm/year. In a billion year's time, the average will be 1.9 cm/year. This is a measure of the long-term trend in plate tectonics, although there will be interruptions by the deep mantle plumes, mentioned again below, that convey core heat to the surface, with environmental impacts discussed in Chapter 9 and raised as a matter of concern in Chapter 15.

Effects of core heat and convective plumes in the mantle

In Chapter 4 we refer to the Earth's magnetic field as an important environmental feature and conclude that, in spite of its uniqueness in the inner solar system, it is robust and will continue indefinitely. This means that the core will continue to lose heat to the base of the mantle, maintaining the buoyancy that drives deep mantle convective plumes. Like everything else in mantle convection, the plumes are not permanent features; they die and new ones appear, as illustrated in Fig. 9.2. When a new plume reaches the surface, it releases a very large volume of basaltic lava, with environmental consequences that are a subject of Chapter 9. Thus, continued loss of heat by the core raises the spectre of repeated episodes of flood basalt volcanism. Although the plumes responsible for flood basalts are driven independently of the plate tectonic mechanism [5.3], they influence it. The reason for this can be seen in the manner of convective cooling of the mantle. The subducting slabs are about 100 km thick, but the subduction zones are separated from one another by thousands of kilometres. The diffusion of heat is so slow that coolness from the thin slabs cannot spread through the deep mantle, even in billions of years, and the same would be true for the scattering of fragments of slabs if they were to break up. Particularly in the case of the lower mantle, which is more viscous than the upper mantle, it is evident that it cannot be cooled by a fixed pattern of convection. The arrangement of subduction zones must repeatedly change, so that different regions of the mantle are cooled sequentially and not simultaneously. Convincing evidence that, at least in many cases, the pattern changes that are seen in the geological record were triggered by the arrival at the surface of new plume heads is presented by V. Courtillot [9.3]. When they push up into the lithosphere they cause it to split and may open up new oceans, causing rearrangement of plate motions and boundaries. An interesting illustration of the consequences of such a change is seen in the Hawaiian Islands.

The Hawaiian Islands are a surface expression of a deep mantle plume that has remained essentially fixed in position at 19°N as the Pacific plate, on which the islands sit, drifts north-west. Hawaii Island itself ('the big island', with currently active volcanoes) is the youngest of the chain of islands, which have ages increasing in the north-westerly direction. The directions of the magnetizations of their rocks show that they were all formed at 19°N, the site of the plume, which has repeatedly punctured the plate as the earlier volcanic edifices drifted away and eroded. The next island in the chain, already named Loihi, has not yet reached the sea surface, but is developing south-east of Hawaii. The chain of islands, many now totally submerged, extends as far as Midway Island, which was formed 43 million years ago. At that point there is a change in orientation, with the line continuing almost due north as the Emperor Sea Mounts, a striking feature of the sea floor topography in Fig. 5.2. This direction change identifies 43 million years ago as the date of a plate rearrangement. Not all plate rearrangements are as well dated, but the identification of new plume heads as triggers for them suggests that they tend to occur at intervals coinciding with flood basalts. This

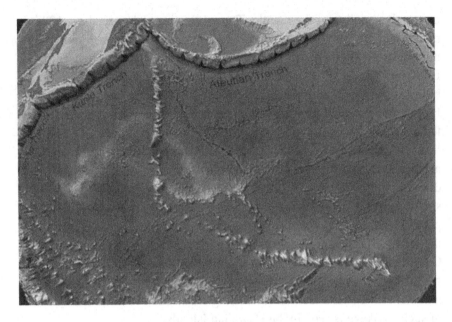

Fig. 5.2. Pacific sea floor topography, showing the hot spot trace marked by the Hawaii Islands and Emperor Sea Mounts. Courtesy of David Sandwell.

certainly happened in several cases, but the time scale for developing thermal instabilities in the mantle, about 50 million years, is longer than the average interval between flood basalts, so no more than a partial correlation could be expected. Although some details of their behaviour are not yet fully understood, plumes and flood basalts are an inevitable feature of global tectonics with a convecting core that generates a magnetic field. The consequences are discussed in Chapter 9 and referred to again in Chapter 15.

Tectonic effects will continue indefinitely with little change

The slowing mantle convection prompts a comment on what this means for earthquakes and volcanic eruptions. A simple argument [5.4] leads to the conclusion that the average plate size remains more or less constant as the plate speed decreases. This means increasing age and thickness of the cool subducting slabs and therefore an increasing volume of material that is cool, in the sense of having a temperature below half of the melting point, and so prone to earthquakes. This largely compensates for the slower plate speeds and no overall decrease in earthquake activity can be expected. With slower plates and undiminished erosion of the continents, marine sediments will thicken, delivering a constant flux of sediment to subduction zones, along with the thickened slabs. Although it appears counterintuitive, on this basis we see no significant variation in subduction zone volcanism as convection slows, with tectonics continuing without major change for billions of years.

The continent-ocean structure

A basic observation that demands an explanation is the clear division of the surface into continental and ocean basin areas (Fig. 5.3). The continents stand higher because they are composed of light, silica-rich rocks and are gravitationally balanced (the principle of isostasy, discussed in Chapter 6), but they must be maintained, because they are being continuously eroded. The world's rivers carry 20 billion tonnes $(2 \times 10^{13}$ kg) per year of suspended sediment. While much of it is deposited on lowlands, coastal wetlands and submarine continental margins and so is not removed from the continents, 20% or so, that is 4 billion tonnes $(4 \times 10^{12}$ kg) per year is carried to the deep oceans. As it accumulates on the ocean floors, it is swept to subduction zones and convectively recycled, but this means that the continental areas above sea level, which are subject to erosion, are progressively worn down to sea level and must be maintained by the recycling process. The total mass of continental rock above sea level is 0.3 billion billion tonnes $(3 \times 10^{20}$ kg) and this much is eroded away, and eventually subducted in 75 million years. The continents are not disappearing quite as fast as this because, as material is removed, the remaining continental mass rises to maintain isostatic balance but, nevertheless, the time scale on which continental material is convectively recycled is not very different from the 90 million

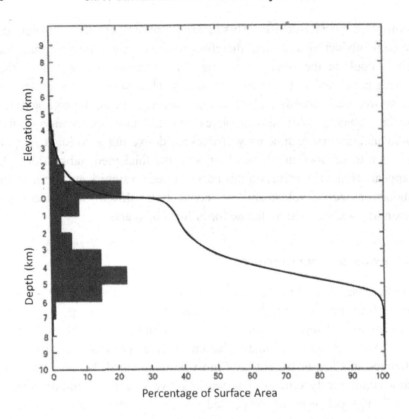

Fig. 5.3. The hypsographic curve shows the distribution in elevation of the Earth's solid surface. Also shown is a histogram of the areas in 1 km elevation intervals.

year lifetime of the ocean floors. We see in this the reason why the land areas are mostly close to sea level. Higher material must have been emplaced by geologically recent or current volcanism, or by uplift resulting from convective compression of the surface plates and it erodes quickly. Not only is convection maintaining the fertility of the land by continuing surface renewal, but without it all land would have been eroded down to sea level long ago. Extensive land areas close to sea level are not vulnerable to erosion and have survived for billions of years on all of the continents. Rapid erosion occurs on the elevated land produced by recent and current tectonic activity.

Planetary evolution involved chemical separation

As mentioned in Chapter 1, the basic structure of the Earth was established very soon after its formation. However, subsequent evolution, driven by mantle convection, has been crucial to life and to development of the resources that have permitted a technological civilization. Our chemical environment is a highly refined selection of the elements, very different from the universe as a whole, or the Sun (Table 5.3), and not even representative of the bulk of the Earth (Table 5.4). If we were to start with the composition of the Sun and extract the elements needed to make the Earth, the starting mass would need to be 200 times the mass of the Earth and if we were to consider just the crust the factor would be much larger still. Nevertheless, the proportions of the major elements in the Earth are similar to their relative proportions in the Sun and the big difference is the rejection of volatile materials, especially the lightest gases, hydrogen and helium, which escape from the Earth's gravity. Our mix of elements began billions of years ago with several stellar processes that produced a cloud of gas and dust, culminating in the injection of debris from a supernova that caused collapse of the cloud to a disc shape, from which the solar system developed. Turbulence caused by the supernova shock wave would have ensured that the cloud was initially well mixed and the sorting processes that led to the diversity of planets occurred during solar system development. Different as they are, the Sun and all the planets had a common origin in a homogeneous nebular soup. The differences are consequences of the processes of planetary accretion, with internal redistribution of material as planets evolved. The evolution of the Earth produced an amazing diversity of surface structures and compositions. The other planets appear simpler and the crucial difference is the existence of the oceans.

Tables 5.3 and 5.4 compare the compositions of the Earth and Sun and also different parts of the Earth. Not represented in these tables are several minor constituents that are strongly concentrated in the crust. As mentioned in Chapter 1, they include the important heat-producing radioactive elements. We should really count the oceans as part of the

Table 5.3. Relative abundances (by mass) of major chemical elements in the Sun and the Earth, using silicon as the reference standard for each.

Element	Sun	Earth
Hydrogen	1003	0.002
Helium	392	10ppb
Oxygen	13.6	2.22
Carbon	4.4	200ppm
Neon	3.5	0.13ppb
Iron	2.6	2.14
Nitrogen	1.6	20ppm
Silicon	1	1
Magnesium	0.91	1.09
Sulphur	0.52	0.2
Argon	0.13	20ppb
Nickel	0.105	0.16
Calcium	0.092	0.12
Aluminium	0.080	0.11
Sodium	0.049	0.013

ppm, parts per million; ppb, parts per billion

crust for this purpose because the concentration of water at the surface is particularly significant. The total water content of the Earth is at least twice the water in the oceans, but that still makes it a very small fraction of the Earth. It is even less abundant on the other solid planets, but, looking at the composition of the Sun in Table 5.3, we see that, apart from the inert gas helium, the components of water (H_2O), hydrogen and oxygen, are dominant. Primitive meteorites contain a lot of water and its rarity in the inner solar system is a consequence of its volatility. The same explanation applies to the next most abundant element in the Sun, carbon. Although, as an element, carbon is not volatile, its compounds with hydrogen and oxygen, methane (CH_4), carbon monoxide (CO) and carbon dioxide (CO_2), are, and these volatile compounds were swept out to the giant planets (and beyond them) during solar system development. Noting that carbon, hydrogen and oxygen are basic ingredients of biological life, it is crucial that the Earth's gravity suffices to hold enough of their volatile compounds to allow the development of biological life and that internal processes have concentrated them at the surface. They are continuously cycled between the atmosphere, ocean and crust and this is especially obvious in the case of water; a volume equal to the oceans is cycled through the atmosphere by evaporation and rainfall in about 3000 years.

Table 5.4. A comparison of the average composition of the upper continental crust with inner parts of the Earth (percentages by mass).

Element	Crust	Mantle	Core
Oxygen	46.8	44.2	5.0
Iron	3.5	6.3	79.0
Magnesium	1.3	22.8	–
Silicon	30.8	21.0	–
Nickel	0.1	0.2	6.3
Sulphur	3.0	0.03	8.6
Calcium	3.0	2.53	–
Aluminium	8.0	2.35	–
Sodium	2.9	0.27	–
Others	3.7	0.3	1.1

The primitive atmosphere

Looking at the abundance of oxygen in Tables 5.3 and 5.4, one may express surprise that our oxygen-rich atmosphere should require a special explanation. Once again, the essential word is 'volatility'. The minerals of the mantle and crust are almost all mixed oxides of silicon, magnesium, iron, aluminium and other elements that bond strongly to oxygen, making much of it securely non-volatile. The other common element with a strong affinity for oxygen is carbon, but the volatility of its oxides and its hydride, methane, has made its terrestrial abundance modest. The restriction on the Earth's oxygen content does not arise from a limited abundance in the solar system as a whole. In the Sun it is more abundant, relative to silicon, magnesium etc., than it is in the Earth (Table 5.3). We can understand the early atmosphere by considering the manner in which the Earth and other solid planets accreted, using as clues the meteorites, which are fragments of asteroids with similar chemistry. Initial condensation from the nebula produced beads of the most refractory (least volatile) materials, which can still be seen in primitive meteorites. Further aggregation into bodies of increasing size (planetesimals) would have incorporated volatile materials within the solids, but would not have led to anything big enough to hold an atmosphere, at our distance from the Sun, until they reached at least 5% of the Earth's mass (corresponding to a sphere with 40% of the Earth's radius).

Harry Hess, 1906–1969
A geologist with a strong connection to the US navy. As a ship's captain in the Pacific in the second world war, he used newly developed sonar equipment to explore sea floor topography. This background led him to the concept of sea floor spreading as the surface expression of mantle convection, providing a mechanism for continental drift, which was being demonstrated by paleomagnetism.

Amalgamation of large planetesimals to produce planets would have involved violent collisions, causing some vaporisation and release of trapped gas. We can get an idea of the likely composition of a very early atmosphere by considering the volatiles that would be released by pulverising and heating a representative collection of meteorites. Water, carbon dioxide and sulphur dioxide would be good candidates (as in volcanic gases), but not free oxygen. Rather, the process would consume any available oxygen, just as volcanic gases and the weathering of crustal rocks do now. The oxygen came later. We note also that the process would have produced no significant methane, which did not accrete in the meteorites or, by inference in the inner solar system, in spite of its abundance in the outer solar system. Also it is never more than a minor constituent of volcanic gases. This is an important consideration in the discussion of atmospheric oxygen in Chapter 7.

The role of biological life

Although the basic chemistry and physics of the crust, oceans and atmosphere were established by the inorganic processes of solar system and Earth formation and by mantle convection, this was only the starting point for development of our environment as it now is. With the necessary conditions in place, biological life took over and made its own environment. The organic processes by which this occurred differ in a fundamental way from the inorganic processes that preceded (and still accompany) them. Whereas inorganic processes tend towards chemical (and thermodynamic) equilibrium, biological processes develop chemical structures that are not in equilibrium with their surroundings. A simple example of the evidence for this is the existence of methane in our oxygen-rich atmosphere. Methane (CH_4) is a simple molecule composed of common elements, carbon and hydrogen, and is abundant in the outer solar system, where it certainly does not have an organic origin, but even the trace in our atmosphere is seriously out of equilibrium with the free oxygen and is a decay product of organic material. It survives in the atmosphere for about 10 years and its coexistence with oxygen in a planetary atmosphere is a sure indication of biological life. The abundant oxygen is central to the features that make the Earth unique and the reason for it requires close scrutiny. The favoured explanation is that it is a biological product resulting from, and maintained by, photosynthesis, the process by which plants use the energy of sunlight to extract the carbon from carbon dioxide, building it into their own structures and releasing the oxygen. While this is true, it is only part of the truth and the relative importance of a simple inorganic mechanism of oxygen production is discussed in Chapter 7. But the abundance of free oxygen is a special situation. It is unique in the solar system and has existed for little more than 10% of life of the Earth.

Most plant material decomposes without burial, reversing the photosynthetic process by consuming oxygen as it does so and we do the same by using fossil fuel, re-establishing chemical equilibrium by returning the carbon to the atmosphere as CO_2. However, vigorous as it is, photosynthesis, per se, is not a sufficient explanation for the oxygen atmosphere. The net production of oxygen requires permanent burial in the crust of the organically fixed carbon. Localised concentrations of it have endowed us with the exploitable fossil fuels, coal, oil and natural gas, but the total quantity needs to be much greater than that if it is to explain the atmospheric oxygen (Chapter 13). The destruction of tropical rainforest is sometimes claimed to be eliminating the mechanism for maintenance of atmospheric oxygen, but that claim needs careful qualification. If left alone most rainforests are CO_2-neutral. They absorb CO_2 and release oxygen as the plant material develops, but reverse the process as it decomposes, and the only net effect arises from permanent burial of the organic material, which may be negligible. But trees are a store of sequestered carbon, typically 6 kg per square metre of mature forest, and this is returned to the atmosphere when trees are destroyed. Cycled plantation forests store, on average, less than half of the carbon in mature forest. Re-establishment of mature forest is slow, but it extracts CO_2 from the atmosphere in the long term. The world's forested area is about 27 million square kilometres, with trees representing carbon storage of 160 billion tonnes (1.6×10^{14} kg). This is fortuitously close to the carbon added to the atmosphere during the industrial era, and about half of the total released carbon when solution in the oceans is included. The total carbon in all living biomass is 3 or 4 times as much.

If we explain the atmospheric oxygen as an entirely photosynthetic product, we are assuming that the buried organic carbon (fossil fuels and presumed widely distributed more dilute forms) is sufficiently voluminous to explain not only the present atmospheric oxygen, but much more. Oxygen is continuously consumed by volcanic gases and the weathering of fresh crustal rocks, so that the total production of oxygen over the life of the Earth must have been many times the present atmospheric content. As discussed in Chapter 13, we have no evidence that the Earth contains that much fossil carbon in a reduced state (that is discounting carbonate rock in which it is not separated from oxygen). No doubt that will be debated, but Chapter 7 presents an alternative mechanism for the production of free oxygen that would have maintained a 'base load supply'. Photosynthesis has become the dominant source only for the last 10% or 15% of the life of the Earth.

6. The oceans

We cannot understand the environment without understanding the oceans

The message that the oceans have a central role in controlling the environment is emphasised in several chapters. At least some ocean formed very early (Chapter 1) and may even have started forming before accretion of the Earth was complete. Its subsequent evolution, with continental crust distinct from the ocean basins, is a result of tectonic processes that are lubricated by ocean water (Chapter 5). Two processes have produced the oxygen-rich atmosphere (Chapter 7) and both depend on an abundance of surface water. The oceans hold vital stores of heat (Chapter 8 and 15) and carbon dioxide (Chapter 13) and have had a controlling influence on development of mineral deposits, both inorganic (Chapter 11) and organic (Chapter 12). These things are well understood, but there are two things that are not as well understood and they are discussed in this chapter: evidence of a resonance in ocean tides that is dramatically changing the tidal effect on the Earth's rotation, and the control of salinity.

Continents and oceans have different deep structures

71% of the Earth's surface is covered by water and we think of this as comprising the oceans, but the 29% that is land is not the full extent of the continents. The continental crust occupies 39% of the area, with almost a quarter of it submerged as continental shelves. The deep ocean basins occupy 61% of the area. This distribution of continental and oceanic crust is obvious in Fig. 5.3, which shows that the solid surface is mostly either close to sea level (the continents) or 3 to 6 km lower (ocean basins). The limited area of intermediate depths is attributable to the steep continental boundary slopes and mid-ocean ridges that mark mantle convective upwelling. A world gravity map (Fig. 6.1) gives no indication of this distinction. There are several ways of displaying gravity variations and the one used in this figure shows undulations of a conceptual surface of constant gravitational potential (energy of gravitational attraction to the Earth), known as the geoid. It can be visualised as mean sea level, with extensions into the continents as the water level in imaginary narrow canals. Its undulations are departures from a perfectly smooth ellipsoidal shape and arise from variations in density within the Earth. The fact that these undulations do not reflect the distribution of continents and oceans demonstrates that the density contrast between continental rock and sea water is compensated by an opposite density contrast at depth. This is a global scale illustration of the hydrostatic balance of the crust, referred to in geology as the principle of isostasy and mentioned in Chapter 5 as a feature of the crustal structure.

Isostasy was first recognised in the 1850s from a survey across north India, which showed that the Himalayan mountain range had much less effect on gravity in the region (as observed by astronomical observations of deflections of the vertical) than would be expected if it were simply superimposed on an otherwise uniform crust. In his interpretation of the Indian survey data, the then British astronomer-royal, George Airy, argued that the crust is effectively floating on the denser mantle in the manner of logs in water. A thick log extends higher out of the water but it also extends deeper than a thin one. Mountains have deep crustal roots (Fig. 6.2) and the same principle applies to whole continents, which is why they do not show up on a gravity map such as Fig. 6.1. The continental crust, of average thickness 37 km (98% of it below sea level) is gravitationally compensated by the denser mantle in oceanic areas), allowing for the much lower density of the sea water itself.

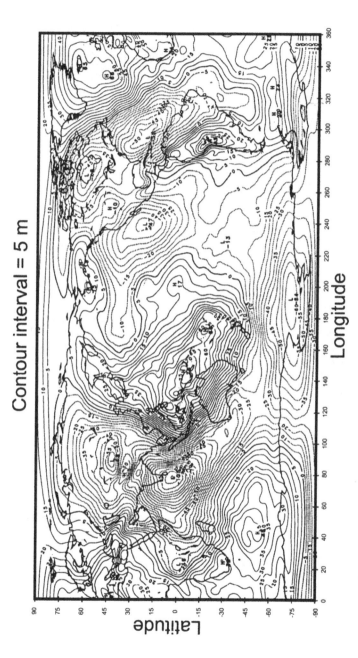

Fig. 6.1. Undulations of the geoid (surface of constant gravitational potential, approximated by mean sea level) caused by internal distribution of mass within the Earth. The fact that highs and lows do not correspond to the large surface features, such as continents and oceans, illustrates the principle of isostasy, whereby these features are effectively floating on the denser mantle. Figure by A. Cazenave, in the volume referenced in note [6.5].

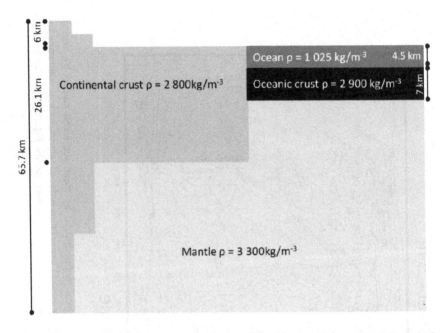

Fig. 6.2. Isostasy according to the 1850s ideas of G. B. Airy, which have stood the test of time. Most of the continental crust is close to sea level (either slightly above or below) and is much thicker than the oceanic crust. Mountainous areas may be up to 6 km higher and have correspondingly deeper 'roots'. Although the assumption of uniform densities of the various layers makes this a simplified model, more realistic models add complexity but no additional insight.

The continental thickness is self-adjusted according to the total volumes of crust and ocean water, so that the surface is close to sea level, maintained by the competition between erosion and convective rejuvenation of crustal material, as discussed in Chapter 5. Thus, the oceans can be seen as filling in the spaces between continents, with a distribution controlled by mantle convection, but it is a mutual control, because the subduction of sea water is essential to the Earth's convective style and because erosion down to sea level controls the continental thickness.

Tides and tidal dissipation

In Chapter 2 we draw attention to the significance of tides in the evolution of the Earth. Tides in the oceans have always been obvious to those who live near them, but the fact that the whole Earth is tidally deformed was not recognised until the 1800s. However, observations of inverted tides in wells were known even before the ancient Greek tidal investigations. Water levels in wells rise and fall as the level in nearby sea does the reverse and this is now recognised as a response to tidal stress in the solid Earth. The fundamental interest in tides that is addressed in Chapter 2 arises from the fact that they dissipate rotational energy. Both solid Earth and ocean tides are involved in this, but it is the marine tide that accounts for most of the dissipation. Thus, a solution to the paradox presented by Fig. 2.3, the dramatic enhancement of tidal dissipation in the geologically recent past, must be sought in the oceans. We can present the observations in a way that makes the nature of the problem more obvious. The dissipation arises from the lag of the tide, relative to the tidal force of the Moon, or Sun, represented by the angle δ in Fig. 2.2. We need to consider how this phase lag can be affected by changes in the oceans.

As discussed in Chapter 2, the effect of the gravitational force of the Moon on the misaligned tidal bulge is to exert a torque on the Earth, slowing its rotation and causing the Moon's orbit to expand. The magnitude of the torque is proportional to the product of the lag angle, δ and the amplitude of the tide, which depends on the masses of the bodies, the distance of the Moon and a numerical factor, k, the ratio of the tidal amplitude to the strength of the tidal force [6.1]. k is a measure of the compliance of the Earth to the deforming force. Although both δ and k are well measured at the present time, both would have varied in the past because the ocean geometry has changed. While there is no record of past ocean structure adequate to infer details of tidal friction and the consequent evolution of the lunar orbit and Earth rotation, they simply respond to variations in the product ($k\delta$) and there is geological evidence of that. Records of tidal layering in sedimentary rocks at two times, 0.62 and 2.45 billion years ago, give the rotation rates at those times [2.3].

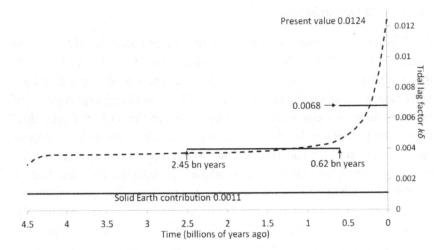

Fig. 6.3. The variation with time of the product (*k*δ) of amplitude and phase lag of the global tide. A constant value would give the Earth-Moon distance shown by the broken curve of Fig. 2.3, implying a dramatic encounter between the Earth and Moon 1.6 billion years ago, which certainly did not happen. In the distant past (*k*δ) was much smaller than the present value. Average values over two time intervals, shown as the horizontal lines, were obtained from observations of tidal rhythms in sedimentary rocks and the broken line in the figure is an empirical fit to these constraints, consistent with the solid curve in Fig. 2.3.

Combined with the present rotation rate, these observations provide average values of (*k*δ) over the two time intervals, 0 to 0.62 billion years and 0.62 to 2.45 billion years ago, which are plotted as horizontal lines in Fig. 6.3. The broken line in Fig. 6.3 is drawn to represent a variation of (*k*δ) that meets the present day value and satisfies the two time averages. Without more data there is no way of deciding just how this line should be drawn, but a simple plausibility argument imposes restrictions. Since the value of (*k*δ) is attributed primarily to the oceans, major changes to it require big changes in ocean basin geometry and these cannot happen quickly. The broken line must be a smooth curve. This leaves little alternative to the conclusion that, over most of geological time, (*k*δ) has been little more than a third of its present value. In some important way, the present ocean basin geometry differs from earlier configurations and the change is recent when viewed on the time scale of Fig. 6.3. The 1557 suggestion of Scaliger, referred to in the second paragraph of Chapter 2, that the Atlantic Ocean is resonant, now appears relevant, because on the time scale of Figs. 2.3 and 6.3, the ocean has opened, increasing the tidal amplitude, and therefore k. We examine the suggestion more closely.

Tidal resonances

The natural (undriven, free) speed of the tide in water of depth h is $\sqrt{(gh)}$, where g is gravity. It is an example of what are colloquially referred to as 'root-g-h waves', a necessary condition for which is a wavelength much greater than water depth. Tsunamis are another example. The Japanese word 'tsunami' has displaced the once common term 'tidal wave' because tsunamis were seen to be unrelated to tides. However, they propagate in the same manner and the term is not at all illogical. The literal translation of 'tsunami' is 'harbour wave' and this is interesting in the present context because tsunamis may be amplified by resonance in harbours, just as, we suggest, the semi-diurnal (12.4 hour) tide is amplified in the Atlantic Ocean. In a geometrically simple case, illustrated in Fig. 6.4, resonance occurs when the time taken by a wave to travel at the free speed across the trough and back coincides with the timing of an oscillatory driving force (lunar gravity or an arriving tsunami) [6.2]. The resonant wave period is controlled by the width and depth of a trough, with a simple relationship between these dimensions for any particular wave period, as in the figure, and resonance occurs when the period of the driving force happens to match the required conditions. Of course, neither the Atlantic Ocean nor Japanese harbours have the neatly simple geometry of Fig. 6.4, which would give sharp resonances. They have irregular outlines and sea floor topography, with consequent broad resonances, that is, some resonant amplification over a wider range of periods.

Local resonances in bays, estuaries and channels occur at tidal frequency as well as tsunami frequencies and they are important to tidal

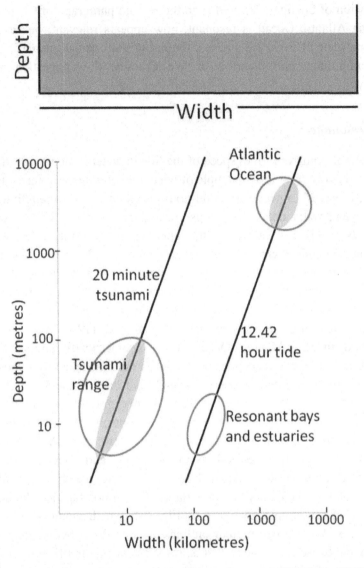

Fig. 6.4. Resonance of tides and tsunamis. This requires a specific relationship between depth and width of a marine basin, according to the period of oscillation [6.2], shown here for the 12.4-hour tide and a typical tsunami period of 20 minutes, for a geometrically simple structure, as illustrated. The dimensions of the Atlantic Ocean are seen to be within the range that makes it tidally resonant.

dissipation because the water motion is rapid. Impressive examples are the Bay of Fundy, eastern Canada, the Bristol Channel, west England, and the Rance estuary on the Atlantic coast of France. It is significant that these examples all border the Atlantic Ocean, because their tides are not driven independently by the Moon (or Sun), but by the ocean tide. Although the dissipation of tidal energy is concentrated in the shallow, marginal seas, it occurs because it is driven by the ocean tide. The shallow seas draw their tidal energy from the ocean tide, which happens to be strong in the Atlantic, and not directly from the Moon's gravity. So, we attribute the dramatic increase in tidal dissipation, apparent in Fig. 6.3, to an increase in amplitude of the Atlantic tide as plate tectonic motion over the last 200 million years or so first separated Europe and Africa from the Americas and then widened the resulting ocean, making its tide resonant.

Development of the Atlantic Ocean as a consequence of plate tectonics

The demonstration of the opening of the Atlantic Ocean, and its timing, were one of the early successes of the study of paleomagnetism. Figure 6.5 shows a sequence of positions of the North Pole, as inferred from the magnetism of rocks in North America and Europe, over a 250 million year period, from 500 million years ago to 250 million years ago.

All magnetic observations agree that there was only ever one magnetic north pole and not different poles for different continents. Thus, the two separated apparent pole paths in Fig. 6.5 provide a measure of the relative motion of the continents. The two curves can be made to coincide if the relative positions of the continents are changed to close the Atlantic, demonstrating that, over that time interval, there was no Atlantic Ocean and that it has opened since. The idea that the coastlines of the continents now bordering the ocean match, and that they were once joined, has a long history, identified particularly with the name of Alfred Wegener, and a modern representation of it is shown in Fig. 6.6. Is such a development of a tidally resonant ocean unique in the

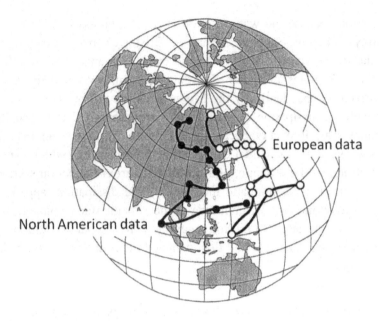

Fig. 6.5. Apparent positions of the north pole from magnetic measurements on European and North American rocks dated from 500 million years ago to 250 million years ago, plotted on a present-day globe. The two pole paths would coincide if, over that time period, the continents were joined in the manner of Fig. 6.6 and have since separated, with the Atlantic Ocean between them. The width of the Atlantic Ocean is still increasing. Figure redrawn from Van der Voo [6.3].

geological record? What reason is there to suppose that the Atlantic resonance is a one-off event? Geometrically it is a very special situation. The ocean extends north-south for 90° (10,000 km), straddling the equator, with a more or less uniform width. The north-south orientation and roughly uniform width are crucial because the tides drive east-west motion of the water and the whole length approaches resonance at the same time, resulting in a major increase in the global coefficient, k, and consequent tidal dissipation. Regarding plate tectonics as a random process and the present Atlantic geometry as a chance event, how often would an ocean with this structure emerge? Although the Atlantic resonance story is still far from a complete, detailed theory, it is the only current hypothesis that can account for an effect as big as that seen in Fig. 6.3.

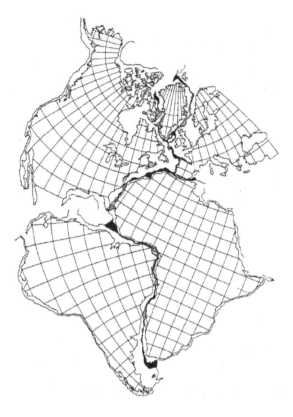

Fig. 6.6. A fit of the Americas to Europe and Africa, with both North and South Atlantic Oceans closed. The continental boundaries are taken at the 500 fathom (914 m) depth contour, to include submerged continental margins. Shaded areas are gaps and black areas are overlaps. Figure redrawn from Bullard et al [6.4].

Other tidal effects

We can add two environmental footnotes to our discussion of marine tides. The repeated exposure and inundation of the inter-tidal zone, the strip of land between the low and high water lines of the coastlines, has diverse life forms, including vegetation adapted to this cycle, living for hours, or longer, alternately either immersed or drying out in air. Was

Alexander von Humboldt, 1769–1859

A German naturalist and explorer with a very wide range of scientific interests. Following detailed study in South America, he became an early proponent of the idea that the Atlantic Ocean was once closed, with South America and Africa joined. His name is remembered in numerous geographical features, including the Humboldt current, off the west coast of South America. Portrait by Gemälde von Joseph Stieler.

Alfred Wegener, 1880–1930

German meteorologist and arctic explorer. A strong advocate of continental drift, he was inspired by the matches in geological features on continental margins now separated by oceans, especially the Atlantic. Wegener had rather few supporters in his lifetime, but his basic ideas are now vindicated, not least by Fig. 6.6, although the rate of continental movement is far slower than he proposed.

this zone a stepping stone for colonisation of the land by plants? The prolonged delay between development of life at sea and on land (at least 1.5 billion years and perhaps 3 billion years) needs a convincing explanation. Chapter 7 makes the suggestion that ozone is the clue. Another important environmental effect of tides is the mixing of ocean water, with redistribution of solutes, including oxygen and carbon dioxide. The salt content acts as a tracer of the mixing process and Fig. 6.7b illustrates the relative ineffectiveness of mixing in the Caspian Sea, which has negligible tides.

The origin of ocean salt: evidence from the Caspian Sea

The salt content of the oceans requires some explanation. The surface salinity observed by satellite, shown in Fig. 6.7(a), illustrates the variations arising from the competition between the increase caused by evaporation and decreases due to rainfall and river run-off. The detailed picture varies, especially with the seasons, but the general pattern persists, with mixing continuously reducing the differences. The mixing is rapid enough to maintain almost constant relative proportions of the solutes as the total concentration varies, and Table 6.1 lists the average abundances of major constituents. As has been recognised for many years, rivers carry to the sea, in dilute form, the soluble products of erosion and these accumulate as the water evaporates and is recycled, but the ocean chemistry is not explicable simply in terms of this accumulation. A comparison of the composition of ocean salt with inland salt lakes that have no outlets gives an indication of the complexity of the problem. The Dead Sea comes to mind, but we learn more from studies of the Caspian Sea, which is, by far, the largest of the world's salt lakes. Whereas the Dead Sea is essentially saturated with salts, and precipitates out some of its constituents, the salinity of the Caspian Sea is hardly more than a third of the salinity of the oceans (Fig. 6.7b), so we might expect its salt composition to be an accurate reflection of its salt input. However, as Table 6.2 shows, this is not true for the observed input and we need a closer look at what is happening.

First, we note that it is not strictly correct to assume that the Caspian Sea has no outlet. It discharges water into a smaller satellite lake on its eastern boundary, named Kara Bogaz Bay. The Caspian Sea is, at present, 26 m below sea level and Kara Bogaz is 0.5 m lower. The connection is a narrow channel, and that is the meaning of the word Bogaz, so its description as a bay is misleading. Kara Bogaz has no outlet and no other surface input and has accumulated salt to the point of saturation, but it is much smaller than the Caspian Sea and neglect of its salt content does not dramatically upset calculations of the Caspian Sea salt, from which we can draw two conclusions. The salt compositions of the Caspian Sea, its river inputs and Kara Bogaz are all mutually incompatible. They cannot be derived from one another or by mixing.

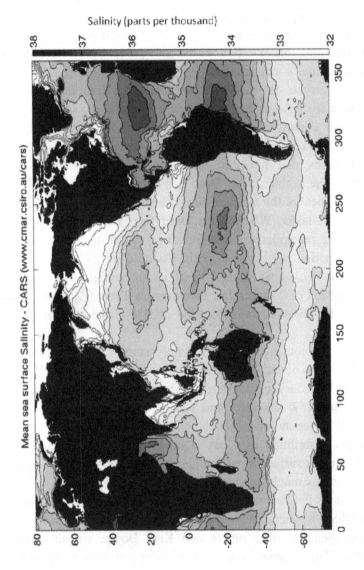

Fig. 6.7. (a) Surface salinity of the oceans. Map by courtesy of Jeff R. Dunn, CSIRO Marine and Atmospheric Research.

Fig 6.7. (b) Salinity of the Caspian Sea. There is a seasonal variation; this is an average for summer. Adapted from The Caspian Environment Programme 2002 (UNEP/GRID-Arendal, Philippe Rekacewicz).

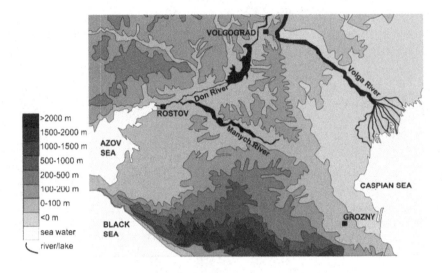

Fig. 6.8. Overflow route for Caspian Sea water during the late stage of the last ice age, 18,000 to 15,000 years ago, when melt water from northern ice caps flowed into it down the swollen Volga River. Both the Caspian Sea and the Black Sea were flushed with fresh water at this time.

The Caspian Sea salt is closer in composition to what is found in the water in some hot springs in the mountainous area near its eastern shore. It is evident that both the Caspian Sea and Kara Bogaz are deriving salt from groundwater. It is inevitable that the same is true for the oceans and we will return to this point presently. The second conclusion to be drawn from the Caspian Sea salt is that if it were entirely derived from the rivers that feed it, it would have accumulated in about 12,000 years. Allowing for the crudeness of this calculation, arising from neglect of groundwater salt and the export of salt to Kara Bogaz, we infer from this number that the Caspian Sea was flushed with fresh water 10,000 to 20,000 years ago. This accords with geological evidence that, at the end of the last ice age, the Volga River carried a large volume of melt water from retreating northern ice caps, with the Baltic Sea still blocked by ice, 18,000 to 15,000 years ago. This caused the Caspian Sea to overflow into the Azov Sea/Black Sea via a low area north of the Caucasus Mountains, known as the Manych Depression (Fig. 6.8). The highest point of this overflow route is about 25 m above sea level, so that flow over it required the Caspian Sea to be 50 m above its present level. The Black Sea would have been flushed at the same time, both by direct flow into it from the north and by the Caspian overflow, and its salinity is still only 2/3 of the ocean salinity, although it has a narrow connection to the Mediterranean, which is saltier than the oceans.

Why are the oceans not salt saturated? The special case of dissolved calcium

The speed with which the Caspian salt has been built up, from a presumed almost fresh start 15,000 years ago, and the fact that undrained lakes may be saturated with salt, pose the question: What limits the concentration of ocean salt? We approach this question by considering first the case of calcium, a component of sea salt that is of particular interest because of its role in fixing, as calcium carbonate (limestone etc.), the carbon dioxide dissolved in the ocean. The calcium in carbonate is about 500 times the calcium now dissolved in the oceans, which is 415 parts per million of the ocean (5.8×10^{17} kg). Rivers are adding about 280 million tonnes (2.8×10^{11} kg) per year, so that, if this were the only input, the oceans would have a 2 million-year accumulation. An immediate concern is that the river input is only 1% of the calcium that would be needed to neutralise (as carbonate) all the carbon dioxide that is produced by fossil fuel burning (Chapter 13). For the moment we can take a longer term view to put these numbers into a global perspective. With the obvious assumption that all the carbonate in the crust originated in the sea, but recognising that some of it includes magnesium as well as calcium, the total mass of calcium required to deposit all of the known carbonate rock is 2.5×10^{20} kg, which is only 20% of the calcium that would be delivered to the sea by rivers in 4.5 billion years (the life of the Earth), at the present rate. Thus, in spite of the fact that much of it is fixed in carbonates, calcium is not isolated from the question of what happens to most of the salt in the sea.

Table 6.1. Abundances of elements in sea water solutes, expressed as parts per million by mass and in moles. Data selected from an extensive table by B. Fegley [6.5].

Element	Form	Abundance (ppm) (Ocean mass = 1.4×10^{21} kg)	Abundance (10^{18} moles)
Cl	Cl^-	19353	764
Na	Na^+	10781	657
Mg	Mg^{2+}	1280	73.7
S	SO_4^{2-}	905	39.5
Ca	Ca^{2+}	415	14.5
K	K^+	399	14.3
C	CO_2, CO_3^{2-}	27	3.15
O	O_2 gas	4.8	0.42
	compounds*	1970	172
Si	SiO_3^{2-}	3.1	0.155
Fe	Fe^{2+}	0.04	0.001
U		0.0032	0.000019

*Sulphate, carbonate, silicate, nitrate

Table 6.2. A comparison of salt lake chemistry with the ocean. Caspian Sea data from N. Clauer et al [6.6] and Red Sea data from P. G. Brewer and D. W. Spencer in [6.7].

| Lake/sea | Caspian Sea system | | | Dead Sea | Ocean system | | |
	Caspian	Kara Bogaz	River input to Caspian[a]		Ocean	River input to ocean	Red Sea hot brine[b]
Na/Mg	4.294	8.154	1.845	0.923	8.423	1.45	120
Ca/Mg	0.460	0.051	3.176	0.401	0.324	3.63	6.5
Cl^-/SO_4^{2-}	1.652	4.768	3.810	453	7.13	0.86	200
Salinity (parts per thousand)	12.0[c]	340[d]	0.227	337[d]	34.5		variable
Level relative to ocean (m)	−26	−26.5	–	−423	0		

[a]Dominated by the Volga River, but with several smaller rivers with different solutes
[b]Representative values with big variations between pools
[c]Average, with strong north-south variation (Fig. 6.7b)
[d]Periodically over-saturated

A hydrothermal source of salt

The discovery of hot, concentrated brine in hollows on the floor of the Red Sea [6.7] drew attention to another source of sea salt. The Red Sea is the early stage of a new ocean, marking the separation of Africa and Arabia. Its median ridge is the spreading centre and lava is emerging to form new ocean crust, as in the other mid-ocean ridges. Circulation of sea water in cracks makes an important contribution to its cooling and the resulting hot water is rich in salts and other dissolved minerals, some of which separate out when they meet the cool sea water. This is an important mechanism for the generation of exploitable deposits, as discussed in Chapter 12. Unlike the open oceans, the Red Sea is almost enclosed, with little water movement, so that, in spite of its high temperature, the brine, being more concentrated, is denser than the cool sea water and flows into hollows on the sea floor. Its existence draws attention to the importance to the composition of sea water of hydrothermal circulation at mid-ocean ridges. It gives at least a partial explanation of the differences between the salt compositions of the oceans and their river input, as illustrated by two examples: (i) In common with its mantle source, ocean ridge basalt is in a strongly reduced (oxygen-deprived) state and converts the sulphate (SO_4^{2-}) ions in sea water to sulphides (FeS, etc.), which promptly precipitate out. (ii) Magnesium from the ocean is absorbed, but calcium in the rock is dissolved and discharged into the ocean. These effects are consistent with the trends in Table 6.2.

Recycling of ocean salt

The salt budget of the river input to the oceans is reasonably well known, but the input by mid-ocean ridges is not. Also unknown is the salt content of groundwater that leaks from the continents into the oceans below sea level. But, these uncertainties do not obscure the fact that the total input is so large that, if it were all retained, the oceans would be salt-saturated several times over. Shallow areas repeatedly filled with salt water and evaporated have left salt layers of various compositions (evaporites) that are widely distributed in the crust. But they are not as voluminous as carbonates, which, as we have seen, can account for no more than 20% of the calcium delivered to the oceans; evaporites represent only a small fraction of the 'missing' ocean salt. But it cannot really be missing in the sense of existing out of geological sight. There would be far too much of it for that. The inevitable conclusion is that the implied quantity of salt does not, in fact, exist. The rivers are not delivering entirely 'new' salt all the time. We infer that it is continuously recycled, returned to the continents with subducted oceanic crust, sediment and sea water. This influences the estimate of dissolved calcium that is available to fix CO_2 as carbonate, a matter that we address presently.

As mentioned in Chapter 5, the subduction of sea water and its re-emergence on the continents received convincing confirmation in a study of a radioactive isotope of beryllium, [10]Be. The remaining problem is to show that this mechanism of salt removal suffices to balance the input to the ocean. It means extraction of at least 10^{12} kg (a billion tonnes) of salts per year, something like 10% of the mass of subducted sediment, far more than could be attributed to salt dissolved in pore water. Deep penetration of the underlying basaltic ocean crust is indicated, with chemical interaction that adds salts to the basaltic composition.

Now we return to the question of the origin of ocean salt, by estimating the relative contributions to it by rivers, groundwater flow and hydrothermal circulation, using the data summary in Table 6.2. We make the assumption, justified presently, that the salt compositions of groundwater and river water are more similar to one another than either is to the salt introduced to the ocean hydrothermally. With this assumption, we effectively have just two different inputs, rivers, with a known composition, and hydrothermal vents, releasing salt presumed to match the composition of Red Sea brine, together producing the well known composition of sea water. This allows us to use the element ratios in Table 6.2 to estimate the proportions of the two inputs. Using the Na/Mg ratios, we find that the hydrothermal input outweighs the river input by a factor of about 3.1 [6.8]. Repeating the calculation with the Cl/SO_4 ratios we obtain a factor of about 3.5. Recognising the uncertainty arising from variability of the input compositions, the reduction of sulphates to sulphides at the mid-ocean ridges and possible modification of sea salt chemistry by interaction with decomposing basalt and sediment, these results agree well enough to indicate a robust, if rough, conclusion: the hydrothermal salt makes by far the bigger contribution. Although we have no separate identification of the groundwater composition, we can use the Caspian Sea data to test the possibility that it seriously biases the results. The salt composition of the river input to the Caspian Sea cannot be made to resemble that of the Sea itself in both Na/Mg and Cl/SO$_4$ ratios by the addition of Red Sea brine. Even with the uncertainty over Caspian Sea groundwater, it cannot resemble the Red Sea brine. Thus, the assumption that the groundwater flowing into the oceans resembles river water is better than the assumption that it resembles the hydrothermal input. This confirms that the composition of ocean salt is dominated by the hydrothermal contribution to it, with smaller contributions by rivers and groundwater flow.

The role of salt in ocean circulation and control of the heat budget

A property of the ocean that has far reaching environmental implications is its heat capacity. The moderation of daily and seasonal temperature cycles in coastal areas is very obvious, but this is a short term effect involving only the shallow parts of the ocean. Longer term consequences of ocean heat storage include protection of the polar caps from rapid melting, as discussed in Chapter 15. To emphasise just how big the ocean heat capacity is, if all of the solar energy arriving at the top of the atmosphere were to be diverted to heating the ocean, bypassing the atmosphere and everything else, with none reflected or re-radiated, and if that energy were uniformly distributed through the entire ocean, its temperature would rise by only 1°C per year [6.9]. Most of the ocean is quite cold (Fig. 6.9) and the low temperature is maintained by cold polar water that sinks to the ocean floor, to be replaced by warmer water from

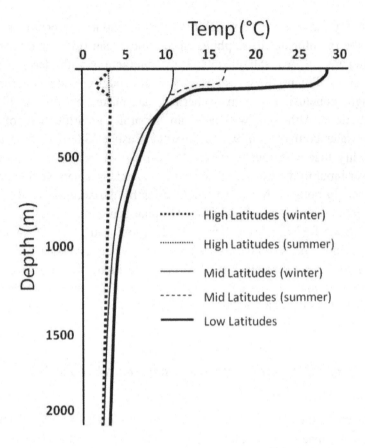

Fig. 6.9. Characteristic ocean temperature profiles for different latitudes and seasons.

lower latitudes. This process is driven more by salinity than by temperature because cold water has a very low thermal expansion coefficient. When ocean water freezes, salt is rejected, producing fresh water ice and leaving an excess of salt in the remaining water, which is driven down by its increased density. This process is referred to as the thermohaline circulation. If the Earth were to warm up enough to stop or seriously limit the winter freezing of polar water, the circulation would be reversed by the high surface salinity in subtropical areas of low precipitation and high evaporation rates (Fig. 6.7a). This was pointed out by T. C. Chamberlin in 1899. It is an extreme prospect, but it draws attention to the sensitivity of the ocean circulation to disturbances that may be less extreme but would have dramatic effects on the redistribution of heat by the oceans. We have in mind the present dominance of the oceans in absorbing the net greenhouse heat input, inhibiting its transfer to the polar caps. This is a transient situation. The ocean temperature rise cannot continue indefinitely. In due course it will approach equilibrium with the atmosphere and the transfer of greenhouse-induced heat to the oceans will slow down or stop. Then atmospheric heat will more effectively reach the polar caps, dramatically accelerating ice cap melting and sea level rise. The mechanism for this is discussed in Chapter 15.

Salinity of the blood of vertebrates

In the late 1800s the salinity of blood, not only of humans but of vertebrates generally, invited a comparison with sea water. Several authors, notably G. Bunge in Germany and R. Quinton in France, concluded that a marine origin for the early ancestors of all of these animals was indicated. In the early 1900s the idea was vigorously pursued by a Canadian, A. R. Macallum, whose documentation was persuasive and made it the conventional view of the origin of land animals, as well as those still living in a marine environment. Absolute proof would be difficult or impossible to obtain and doubts have persisted, but no satisfying alternative explanation of blood salinity has appeared in 100 years and apparently contradictory evidence has not been difficult to dismiss. With the advent of DNA analyses the case for a common distant ancestry of all animals has become compelling and, since life on land was a late starter, compared with marine life, a marine origin for all life forms has become the obvious inference.

Discussions of the differences between the compositions of blood and sea water have focussed on the relative abundances of sodium (Na), magnesium (Mg), calcium (Ca), potassium (K) and chlorine (Cl), as in Table 6.3. The element ratios in blood are somewhat variable between species — and even individuals of a single species according to

environment — but are broadly consistent over the whole range of life
forms, and the numbers in the table are representative. The differences
from the sea water ratios are systematic but small, except for Mg. The
total salt concentration is more variable between species, being
comparable to that in sea water for many marine invertebrates but only
a quarter of that for mammals, although it is higher in embryos than
in fully developed individuals. Plants have significantly different
concentration ratios and for insects the ratios are generally closer to those
of plants than to the mammalian abundances. A supposition that
differences between the compositions of body fluids of animals reflect
variations in ocean composition at the time of their development appears
unsupportable, with a lack of evidence of significant changes in ocean
composition. The variations in body fluid chemistry are better explained
as evolutionary developments, responses to changed environmental
needs and the remarkable thing is that the variations are so slight.

Table 6.3. A comparison of the abundances of elements in the sea and in mammalian
blood. Values are relative to the abundance of sodium.

Element	Ocean	Blood
Na	1	1
Mg	0.12	0.007
Ca	0.039	0.031
K	0.036	0.068
Cl	1.81	1.29

Availability of calcium for carbonate production

In the long-term, the calcium (Ca) dissolved in sea water provides the
major limitation on the total atmosphere-ocean CO_2 content and we
consider what is known about this in contemplating its availability for
increased natural sequestration of CO_2 as carbonates. We have
mentioned that the river input of Ca to the oceans (280 million tonnes per

year) is less than 1% of the quantity required to neutralise CO_2 at the present rate of its industrial release. But salt input by hydrothermal circulation exceeds the river input by a factor 3 to 3.5 and, if this factor applies to the Ca content, the total Ca input to the oceans is nearer to a billion tonnes (10^{12} kg) per year. This is still short by factor 30 or so of what would be needed for steady-state sequestration of the industrial CO_2, but that does not fully reflect the Ca-CO_2 balance. The Ca availability is reduced by its subduction, along with other components of sea salt, in the sea floor spreading process. Noting that the total carbonate precipitation accounts for about 20% of the river input of Ca to the oceans in the life of the Earth, and therefore perhaps 6% of the total input, and assuming this process still to be removing Ca from the oceans at the long-term average rate, the estimated availability of Ca for CO_2 sequestration is reduced to about 200 million tonnes per year — hardly more than 0.6% of what would be required for complete neutralisation of all industrial CO_2. In any case it is consumed in fixing the naturally produced CO_2 from volcanoes. As mentioned in Chapter 13, carbonate precipitation is a delicate chemical process and is not simply controlled by the abundances of Ca and CO_2, but the abundance calculation presents an over-riding limitation on what may be considered possible.

7. Planetary atmospheres and the appearance of free oxygen

Atmospheres give clues to planetary evolution

The principal components of the permanent atmospheres of the three solid planets that have atmospheres are listed in Table 7.1. The abundances are given as fractions of the total planet masses to draw attention to the fact that the atmospheric densities are very different. But the compositions are quite similar for Venus and Mars. As pointed out in the discussion of primitive atmospheres in Chapter 5, the original atmospheres of the planets were released from the accreting planetesimals and not by direct accumulation of gases from the solar nebula. The compositions of the inner planets are quite similar and their original atmospheres would also have been similar, probably resembling those of Venus and Mars now. The Earth is the odd one out and its difference from the others presents several clues to the evolution of the terrestrial environment. The Earth is, to our knowledge, the only planet with biological life, but although this has been a major factor in the development of the present environment, it must be regarded as an effect of the basic physical difference from the other planets rather than a primary cause. We need to consider what features of the Earth make it suitable for biological life. In what ways does the Earth differ from Venus and Mars? Obvious differences are that the Earth has oceans, a large satellite (the Moon) and it has a magnetic field. Distance from the Sun is also important. More subtle evidence is seen in the isotopes of atmospheric gases. Evidence obtained from the atmosphere itself is considered in this chapter.

Table 7.1. Abundances of atmospheric constituents of three planets, expressed as parts per billion of total planet mass, with percentages of atmospheric masses in parentheses.

Constituent	Venus	Earth	Mars
Nitrogen, N_2	2278 (2.25 %)	664 (73.7 %)	6.8 (1.77 %)
Oxygen, O_2	~0.3 (3×10^{-4} %)	203 (22.5%)	0.37 (9×10^{-2} %)
Argon, Ar	6.5 (6.4×10^{-4} %)	11.3 (1.25 %)	5.8 (1.5 %)
Carbon dioxide, CO_2	98715 (97.7%)	0.52 (6×10^{-2} %) 10 with oceans	377 (96.7 %)
Sulphur dioxide, SO_2	23 (2.3×10^{-2} %)	0.0000001 (10^{-10} %)	– (0%)
Water, H_2O	3 (3×10^{-3} %)	up to 22 (2.4 %) 235 000 with oceans	up to 0.035 (9×10^{-3} %)
Methane, CH_4	– (0%)	0.015 (1.7×10^{-3} %)	– (0%)
Helium, He	0.11 (1.1×10^{-4} %)	0.00064 (7×10^{-5} %)	– (0%)
CO_2/N_2 ratio	43	0.00078	55
$^{40}Ar/^{36}Ar$ ratio*	1.2	296	3000
D/H ratio[†]	0.032	0.000312	0.0016

*Ratio in the solar wind 0.14

†Deuterium/hydrogen ratio in Jupiter's atmosphere 0.00005

Dense carbon dioxide has come and gone

We have referred to volcanism as an essential factor in the development of the environment and now take a closer look at what that means for the atmosphere. The three most abundant volcanic gases, identified as atmospheric constituents in Table 7.1, are, in order of decreasing abundance in volcanic emissions, H_2O, CO_2 and SO_2. Presumably this was true also for volcanoes on Venus and Mars when they erupted or fumed, but we can see from Table 7.1 that not all of these gases have remained in the atmospheres. Consider first CO_2, which is completely dominant in the atmospheres of both Venus and Mars, but exists only as a trace in our atmosphere, although it is an important trace and, as in the table, we should really include the CO_2 dissolved in the oceans, where it is 50 times as abundant. The rate at which CO_2 is added directly to the atmosphere by volcanoes is about 65 million tonnes per year [7.1]. To estimate what this would amount to if simply accumulated over the life of the Earth, we note that the Earth was more active in the remote past and allow also for submarine volcanism, which is not included in this number. With these adjustments the average CO_2 emission over the life of the Earth is estimated to be about 200 million tonnes/year. Much of this has been recycled by sedimentation and volcanism, but if we ignore that and simply calculate the total emission, it would amount to 150,000 ppb of the Earth's mass. As seen in Table 7.1, this is 50% more than the CO_2 in the Venus atmosphere but, since it ignores the recycling and is therefore an overestimate, the Venus abundance is a reasonable measure of the CO_2 that has been sequestered in the Earth over geological time by processes that do not occur on Venus (or Mars). This is one of the important indicators of the processes that have resulted in our environment. But, to put this number into the context of our present environmental problems, the average rate of volcanic emission is less than 1% of the current rate of CO_2 emission by fossil fuel burning [7.2].

Most of the 'missing' CO_2 is locked up in carbonate rocks, limestone and dolomite. These are calcium, (or calcium + magnesium) carbonates, which are essentially compounds of CO_2 with calcium and magnesium oxides, common components of crustal rocks. Much of the world's limestone is a fossilised accumulation of shells of marine creatures, which have combined CO_2 with calcium, derived from the erosion of continental rocks and dissolved in sea water. There is a sensitive chemical balance in this process, requiring a restricted range of acidity of the sea water, with suitable concentrations of dissolved CO_2 and calcium. Limestone is also deposited inorganically, without a biological agent and this was evidently important early in the life of the Earth, before the development of shelled creatures. As considered in Chapter 6, the rate at which these processes can occur is linked to the rate at which dissolved calcium is produced by continental erosion and by the convective circulation of sea water in the fresh crust at mid-ocean ridges. The massive CO_2 atmosphere of Venus indicates that any ocean it may once have had must have been very short-lived, with negligible opportunity for sequestration of CO_2 as carbonate. But, while these mechanisms are essential to removal of much of the CO_2 from the atmosphere, they do not contribute to the free oxygen. There is no separation of the carbon from the oxygen, which remain combined in the limestone. Oxygen is produced only by quite different processes, which separate it from the carbon in CO_2 and from the hydrogen in water (H_2O).

Oxygen from the loss of hydrogen to space

Photosynthesis as a generator of atmospheric oxygen is referred to at the end of Chapter 5 and, in Chapter 12, we present an estimate of the total amount of oxygen that has been produced in this way. It is central to the whole environment story. The conventional view is that it has been the dominant (or even the only) mechanism for oxygen production, but we now consider another one that has been generally discounted for reasons that we consider unsatisfactory. In the upper atmosphere, water is dissociated by solar ultraviolet radiation, releasing hydrogen, some of which escapes to space, leaving the heavier oxygen gravitationally bound to the Earth. Isotopic evidence of this process, drawn to attention many years ago by H. Urey and colleagues [7.3], is persuasive but has generally been disregarded. We now have additional evidence not available to Urey, on hydrogen isotopes in the stratosphere and on other planets, supporting Urey's contention, which can no longer be ignored [7.4]. The process is complicated by other atomic interactions and by the extensive magnetic field of the Earth, which partially isolates a large volume containing hydrogen (the magnetosphere, see Chapter 4) from interplanetary space, but that does not affect the isotopic evidence. It has been argued that the coldness of the upper atmosphere (stratosphere and mesosphere) restricts the amount of water vapour that it can hold or transmit, preventing sufficient water getting high enough (even as ice crystals) to allow the escape of hydrogen to become significant [7.5]. Observations of noctilucent clouds, which we discuss presently, contradict this argument, but if it is accepted, then the estimated present rate of escape of hydrogen would, if accumulated over the life of the

Earth, leave a mass of oxygen three times the present atmospheric content [7.6], but this is far too little to oxidise all the volcanic gases and to weather crustal rocks. On this basis, photosynthesis has been accepted as the only significant source of atmospheric oxygen, essentially by default and not because it is demonstrably adequate. The assessment in Chapter 12 of the total oxygen released by photosynthesis in the life of the Earth, and not consumed by decomposition of vegetation, indicates that photosynthetic oxygen is insufficient and evidence of the hydrogen loss mechanism is examined in this chapter.

The isotopic evidence of hydrogen loss

Hydrogen has two stable isotopes. An atom of ordinary hydrogen has a nucleus that is a single proton, with an orbital electron giving electrical neutrality. It has a mass of one atomic mass unit. A mass 2 isotope, which has a nucleus consisting of a proton plus a neutron, is known as deuterium and is represented by the symbol D. Harold C. Urey was awarded the 1934 Nobel Prize for chemistry for its discovery and his later work on its natural occurrence led to recognition of the escape of hydrogen from the Earth. Like ordinary hydrogen (protium), deuterium has a single orbital electron, making it chemically the same and it can be represented by the alternative symbol ^2H, to distinguish it from the ordinary ^1H. The factor 2 difference between the masses of these isotopes gives them slightly different properties. Water containing only the light isotope evaporates more easily, so that atmospheric water vapour has a slightly lower proportion of 'heavy water' than does the ocean and this difference extends to rain water and polar ice, which have less deuterium than sea water, as discussed in Chapter 3. Similarly, the light isotope escapes more easily from the top of the atmosphere, so that, over geological time, terrestrial water has slowly become enriched in deuterium. There are other factors that favour escape of the light isotope. It is slightly less tightly bound to the oxygen in water and so is more readily separated by solar ultraviolet radiation. More importantly, as referred to presently, condensation-evaporation cycles in the stratosphere cause selective upward movement of the light isotope [7.7], accounting for its greater rate of escape. We use this observation in estimating (below) the total oxygen released by hydrogen escape.

Isotopes in water from deep in the Earth

The crucial observation on hydrogen loss from the Earth is the low D/H ratio of juvenile water. This means water from very deep in the Earth that has had no evident contact with surface water and may, therefore, approximate the water with which the Earth started. With two Japanese colleagues, Urey reported such observations [7.3], using particularly water extracted from Hawaiian rocks, which are derived from a deep mantle volcanic plume. Their conclusion, repeated by Urey on several occasions, is that juvenile water has a systematically lower deuterium content than surface or ocean water. We dismiss one of their two possible conclusions, that the Earth accreted in layers, starting with the iron core and a low deuterium layer before the surface layers arrived, and accept the conclusion that surface water has lost an amount of the light isotope (protium) equivalent to at least 4% of the hydrogen in the present oceans. For several reasons that estimate must be recognised as a lower bound. It assumes that only protium escapes and it neglects any possible mixing of the juvenile water with buried surface water. Ignoring the mixing question, we can estimate the total hydrogen loss from the variation of the D/H ratio in the stratosphere. This results from the fact that vapour containing deuterium atoms condenses more readily to droplets than ordinary (light) water and the droplets fall, leaving the light water vapour at greater height. Deuterium depletion by up to 90% has been suggested, but the overall average stratospheric depletion, relative to surface water, is nearer to 65% [7.7]. We use this number as a measure of the D/H ratio of the hydrogen that escapes, that is 35% of the surface ratio, to calculate the total mass of escaping hydrogen and hence the oxygen release [7.8]. A 4% increase in the D/H ratio of ocean water by the escape of hydrogen with a D/H ratio that is 35% of the ocean ratio, requires escape of 5.8% of the ocean hydrogen and leaves oxygen amounting to 59 times the present atmospheric oxygen. In Chapter 12 we point out that this outweighs the *net* oxygen production by photosynthesis in the life of the Earth and it is a minimum estimate because the possibility of mixing is not allowed for. The total oxygen released by both mechanisms, 100 times the present atmospheric oxygen, is of the right order for agreement with the estimate of oxygen consumed by weathering that we refer to later in this chapter.

Harold Urey, 1893–1981
American physical chemist; discoverer of deuterium (heavy hydrogen), for which he was awarded the 1934 Nobel Prize in chemistry. He made wide-ranging contributions to chemical and isotopic compositions of the Earth and other astronomical bodies. An experiment conducted jointly with a student, S. L. Miller, produced amino acids, the building blocks of biological life, by an electric discharge in a mixture of inorganic gases [1.13]. Urey repeatedly drew attention to the isotopic evidence of hydrogen escape from the Earth as a source of atmospheric oxygen. Photograph courtesy of NASA.

Hydrogen loss from methane

As a point of clarification, we comment on the apparently alternative possibility of hydrogen escaping from atmospheric methane (CH_4), which is of interest because methane is a non-condensing gas that is unaffected by the coldness of the stratosphere and mesosphere. Methane is a very minor constituent of the atmosphere (Table 7.1) and survives only for about 10 years, its ultimate fate being oxidation to carbon dioxide and water. In the Earth it is produced by methanogens, microorganisms often referred to as methanogenic bacteria, although they are not strictly bacteria but arcaea, which are subtly different. There are several kinds, living in a range of anoxic environments. Some produce methane from decaying vegetation in stagnant bogs, digestive tracts of cattle and in the more deeply buried material that develops into the fossil fuels, as considered in Chapter 12. A particularly interesting variety does not require pre-existing organic material, but lives in the extreme conditions of ocean ridge volcanism, where hot, basaltic rock meets sea water. The basalt is rich in iron in the reduced (low oxidation) state, with divalent ions (Fe^{2+}) that are oxidised by the methanogens to the trivalent state (Fe^{3+}) by extracting oxygen from the water and using the hydrogen to convert carbon dioxide to methane [7.9]. The early atmosphere would have had some methane as soon as methanogens appeared. So, if we postulate that hydrogen escapes from the upper atmosphere after release from dissociated methane, we must consider that as part of a cycle of organic carbon. The remains of the disrupted CH_4 molecules are oxidised to carbon dioxide and water, but these are basic ingredients for photosynthesis (and eventual production of methane), and that process releases oxygen. The net effect is that some hydrogen has escaped, some water is lost and oxygen is left behind, but the carbon has simply completed a cycle. Although in this case the hydrogen loss does not derive directly from photodissociation of water vapour, but is mediated by methane, the effect is the same.

Hydrogen loss from other planets

The deuterium/hydrogen (D/H) ratios listed in Table 7.1 show very wide variations between the planets and we examine them for clues to the hydrogen loss problem. The proportion of deuterium is very low in Jupiter's atmosphere and, because the gravity of Jupiter is strong enough to hold all of its hydrogen, this has been interpreted as the primordial ratio with which the solar system began. However, this cannot be relied on because there are solar system variations in isotopic as well as chemical composition, at least in part related to distance from the Sun. Another common reference for primordial composition is the primitive (carbonaceous chondritic) meteorites, but, again there are difficulties. Some of these meteorites contain grains that originated in the atmospheres of earlier stars and were not homogenised with the rest of the material when the solar nebula formed. Their compositions reflect a variety of processes of element synthesis, with isotopic ratios quite different from the bulk of the solar system. Similarly, we cannot use data on the solar wind, or the Sun itself, because deuterium is involved in the nuclear processes in its interior. So, we are reduced to comparing the Earth with Venus and Mars.

In identifying the high D/H ratios of Venus and Mars with selective loss of the light isotope from dissociated water vapour, we can see how the ratios correlate with the remaining water on these planets. Venus has very little water and is so hot that most of what there is must be in the atmosphere. The hydrogen loss is nearly complete and the very high D/H ratio reflects this. Mars, being cold, has a very modest atmospheric water content, but it has water condensed (and frozen) in its crust. It is not as severely water-depleted as is Venus and its D/H ratio reflects this less complete hydrogen loss. But it is still a severe loss compared with the Earth, which is unique in retaining abundant surface water. Evidence that Mars has lost a great deal of water is seen in surface features identified as consequences of water erosion, which could not have happened with the water it has now. Free oxygen could not have remained for long in the Martian atmosphere and the trace that it has is an indication that the hydrogen loss is ongoing, whereas Venus has an even smaller trace. The Venus and Mars data demonstrate that hydrogen loss from dissociated water vapour is an important planetary process.

Noctilucent clouds

For more than 100 years we have been aware of the occurrence of what are known as noctilucent (night luminous) clouds. Fig. 7.1 is a good example. They are observed in summer time at high latitudes (50° to 70° north and south), when the Sun is well below the horizon but still illuminates the clouds because of their great height (Fig. 7.2). They are observed at heights of 75 km to 85 km, far above the stratosphere (or any ordinary clouds) and near to the upper boundary of the next identified atmospheric layer, the mesosphere. The mesosphere is very cold (as in

Fig. 7.1. Noctilucent clouds photographed near Edmonton, Canada, on 2 July, 2011 by Dave Hughes (NASA).

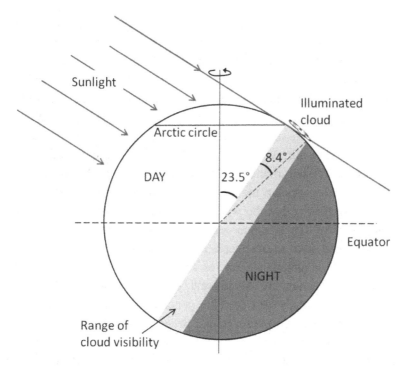

Fig. 7.2. Geometry of noctilucent cloud illumination. This represents the situation in the northern hemisphere mid-summer, when the area north of the Arctic circle (66.5°N) is sunlit for all 24 hours of the day. Cloud at 80 km altitude in the midnight time zone, on the extreme right of the diagram, would be sunlit up to 8.4° south of the Arctic circle by rays that just graze the top of the troposphere at about 10 km altitude. Before and after midnight cloud would be illuminated further south, as indicated by the area of light stippling.

the figure inside the back cover), but above it is the thermosphere, in which temperature increases with height, so noctilucent clouds occur near to the lower boundary of this warmer layer. For many years their nature was a mystery, but more recent work, especially using satellites, has confirmed a speculation that they are composed of ice as very fine crystals [7.10]. Satellite observations also revealed that the clouds observed at ground level are merely the ragged edges of a widespread, brighter polar cloud layer, now known as polar mesospheric clouds. Either the ice crystals form by direct condensation from vapour, implying an abundance of water vapour at altitudes at which that has generally been regarded as impossible, or they are transported from lower altitudes. Although mechanisms for putting the ice so high are subject to continuing investigation, there are at least two possibilities. Vigorous seasonal stirring of the polar stratosphere and mesosphere includes strong vertical winds, which present the most obvious explanation. These are accompanied by temperature changes and appear to be stimulated by ultraviolet radiation from the Sun, especially during periods of solar disturbance. This is consistent with an observed correlation between polar mesospheric clouds and the sunspot cycle. Another possibility is that the water is produced in situ, by oxidation of methane, because that is a non-condensing gas and its transfer through the stratosphere and mesosphere is not inhibited by the coldness. However, its low abundance makes that explanation less convincing. Whatever the mechanism, noctilucent clouds demonstrate that more water reaches the mesosphere than has generally been recognised.

The upper mesosphere, where these clouds occur, is within the lower part of the ionosphere, where solar ultraviolet (UV) radiation breaks atmospheric atoms and molecules into electrically charged components (ions). The mesospheric water is subjected to this bombardment, freeing up some of the hydrogen for escape. Direct evidence that this is happening is provided by satellite observations of a correlation between variations in solar UV radiation and brightness of the mesospheric clouds, which diminishes in response to increased UV intensity, but with a delay that still needs to be explained. The evidence does not suffice for a quantitative estimate of the rate of hydrogen escape resulting from this effect, but it removes what had been accepted as a limitation based on the coldness of the upper atmosphere. An entertaining and informative commentary, presented as a song by NASA staff [7.11], has a repeated punch line emphasising this point:

"Noctilucent cloud ... We didn't think you'd be allowed..."

Does hydrogen loss produce molecular oxygen (O_2) directly?

The low deuterium in juvenile water, the low water contents and high D/H ratios of the Venus and Mars atmospheres, and the existence of noctilucent clouds, have all been known for many years, but have not received the attention they merit. Together they make a forceful argument for hydrogen loss as a mechanism for production of atmospheric oxygen. But the detailed chemical process is not so clear. The disruption of water molecules and release of hydrogen leaves OH^- ions, which are very reactive and form other molecular structures, especially with nitrogen, which is the most abundant atmospheric gas. But these molecules will, like OH^-, be strongly oxidising and if many of them simply mix into the atmosphere and eventually reach the surface, they would oxidise weathering rock and have the same effect as oxygen, without passing through the stage of producing O_2 molecules. But the trace of oxygen in the Martian atmosphere indicates that there is a chemical pathway for production of molecular oxygen from the OH^- ions. In any case only a fraction of the oxygen produced by either photosynthesis or the hydrogen loss mechanism has remained in the atmosphere, and this is even more true for Venus and Mars. We now contemplate its fate.

Oxygen loss

Venus and Mars have lost almost all of the water that we presume they inherited from the solar nebula, and we argue that this was due to photolysis by solar ultraviolet radiation, followed by escape of the hydrogen. In the case of Venus, which is only a little smaller than the Earth, the residual oxygen would not have escaped, but would have been consumed by oxidation of crustal rocks, a process that was probably completed long ago. The situation of Mars is not as clear because, although crustal oxidation would be expected, the gravitational potential at the surface of Mars (the energy per unit mass required for escape) is only 20% of the value for the Earth. This means that atoms can escape from the Martian atmosphere if they are 5 times as massive as the limiting mass for escape from the Earth. Since helium, ^4He, which has 4 atomic mass units, escapes from the Earth, with a time constant of order 100,000 years, atoms of mass 20 would be expected to escape from Mars in the same time. But atomic oxygen, O, has a mass of only 16. Molecular oxygen, O_2, in the upper atmosphere of Mars would be split by UV radiation into separate O atoms or ions and be able to escape. Either way, we would expect little oxygen in the atmosphere of Mars and the fact that it has some indicates that significant hydrogen loss is still in progress.

For an idea of the rate at which oxygen is consumed by volcanic gases and the weathering of crustal rocks on the Earth, we return to the observation in Chapter 5 that 0.3 billion billion tonnes of rock is eroded from the continents in 75 million years and carried to the sea by rivers and streams, a rate of 4 billion tonnes per year. This amounts to about 1.5 cubic kilometres of rock per year, which appears to be a very modest estimate of the average because individual volcanic eruptions can throw up more than this. This much material is replaced by volcanism, either directly or indirectly, thickening the continental crust (and accelerating erosion). Complete weathering of fresh igneous rock consumes oxygen primarily by oxidising the iron (a process that was responsible for the major iron ore deposits, as discussed below), and to a lesser extent aluminium dissolved from the minerals, reducing the rocks to clay (plus some quartz in the case of eroded granite). The oxygen absorbed from the atmosphere may be up to 2% of the mass of the weathered rock. For an estimate of the oxygen removed from the atmosphere by weathering, we take 2% of 4 billion tonnes, that is about 80 million tonnes (8×10^{10} kg), per year. At this rate, the present atmospheric oxygen would be consumed in 15 million years. How representative is this of the distant past? And how is an oxygen supply of this magnitude maintained? The geological record gives some clues.

The history of oxygen development: the record of the banded ironstones

Particularly useful indicators of the availability of oxygen are the iron ores known as Banded Iron Formations, often abbreviated to BIFs by geologists [7.12]. It is significant that they are all very old. Although they are not all identical, BIFs have common features that indicate the conditions under which they formed by chemical deposition on the sea floor in a low oxygen environment. As mentioned in the discussion of methane production at mid-ocean ridges, there are two states of oxidation of iron, termed ferrous for the less oxidized form, with equal numbers of atoms of iron and oxygen, FeO, and ferric for the more oxidized form, in which oxygen atoms outnumber iron atoms 3:2, Fe_2O_3. Ferrous iron is soluble in sea water but ferric iron is not. In the absence of oxygen, ferrous iron remains in solution, but if oxygen is introduced it is absorbed by the iron, which precipitates out in the ferric form. BIFs result from oxidation of iron dissolved in the sea, although they are not all fully oxidised and may include an intermediate oxidation state, magnetite, Fe_3O_4. They are so called because they are sequentially layered, with iron oxide alternating with silicate, in some cases with very thin layers extending for hundreds of kilometres, indicating deposition in calm, very uniform conditions, with cyclic availability of either oxygen or iron (or both). We envisage areas of the sea that are either almost enclosed, and therefore relatively calm, or deep enough to be clear of surface disturbances. They were either rich in dissolved ferrous iron or continuously supplied with it hydrothermally and were periodically presented with an influx of oxygen.

The oxygen that caused precipitation of the banded ironstones must have come from shallow areas of the sea, either by solution from the atmosphere or by in situ photosynthesis by marine algae. By the photosynthetic interpretation one would expect to find a trace of organic carbon in the iron deposits. This has been reported even in very early ironstones, dating from 3.8 billion years ago, and interpreted as evidence of marine life at the time. However, the evidence is disputed. In the cases for which the organic carbon has been subjected to careful examination it was shown to have been introduced to the ironstones much later than their original formation and not subjected to the same metamorphic processes [7.13]. Contrary to what has often been supposed, we conclude that the very early ironstones (and perhaps later ones) absorbed oxygen that had been dissolved from the atmosphere and that it was not produced by photosynthetic bacteria. There was no land life at the time but, with the hydrogen loss mechanism for generation of atmospheric oxygen, there is no problem with an atmospheric source. Then the layering can be identified as a seasonal effect. The exchange of water between shallow and deep parts of the sea is slow because it is thermally layered. Cold, deep water underlies warmer shallow water, which is less dense and so is gravitationally stable, with no tendency for overturn. But this stable situation breaks down when surface water is seasonally cooled and especially if some of it freezes, leaving denser, saline water, which sinks to the sea floor, taking with it oxygen dissolved from the atmosphere, presenting the deeper water with annual pulses of oxygen.

Although some BIFs originated earlier, they are concentrated in the age range 2.6 to 1.8 billion years, with an apparently abrupt termination 1.8 billion years ago. The inference is that, by then, the oceans had become sufficiently oxygenated to precipitate out all the dissolved iron as soon as it appeared and the seasonally layered precipitation stopped. The earliest known banded iron formations are at Isua, west Greenland, where they appear in a sequence of metamorphosed sediments that are associated with igneous rocks dated at 3.77 billion years, and in northern Canada. Older rocks are extremely rare, simply because they have been too completely reprocessed by later geological activity, so it is improbable that banded iron deposits much older than these will be found, even if they once existed. So, there was a trace of oxygen as early as 3.8 billion years ago, but the lack of more abundant ironstones before 2.6 billion years ago indicates that they remained extremely rare for more than a billion years after formation of those in Greenland and Canada and that they may then have been very localised occurrences. But perhaps the 1.2 billion year time gap is more apparent than real and arises because of the rarity of very old rocks. There is also an interesting younger sequence of banded ironstones, with an age range 800 million to 600 million years, another, apparently fortuitous, billion-year time gap from the earlier ironstones. At that time the Earth was heavily glaciated and these younger ironstones are postulated to have formed in stagnant, anoxic water underneath ice shelves that were sufficiently extensive to cover one or more sections of ocean ridge with hydrothermal processes providing dissolved iron to water that was spasmodically injected with oxygen. That interpretation implies that oxygen in the sea was still very limited 600 million years ago.

Control of the oxygen abundance

With the banded ironstone record, we have dates for the progressive increase in free oxygen that allow an interpretation of the factors that controlled it. In the Precambrian period, before 550 million years ago, all life was in the sea and there is no fossil record of land plants. With the land entirely bare, erosion would have been rapid, continuously exposing fresh rock to weathering, with consequent prompt consumption of the available atmospheric oxygen. The oxygen would have been produced continuously, and we presume steadily, by photolysis of water vapour, so its availability in the atmosphere was controlled by the rate of weathering, including the oxidation of volcanic gases. As the Earth cooled, tectonic and volcanic activity gradually decreased and the consequent demand for oxygen diminished, so that by 2.6 billion years ago a little of it was dissolving in the oceans, allowing more rapid precipitation of the ironstones to begin. At 1.8 billion years ago the oxygen supply sufficed to precipitate all of the available iron as soon as it appeared and the cyclic precipitation ceased. However, the abundance of oxygen remained very low until 500 or 600 million years ago, when plant life began to colonise the land. That had two effects. It provided some photosynthetic oxygen, although probably very little initially and, by covering the land, it reduced the erosion and consequent oxygen demand by weathering, allowing an increase in atmospheric oxygen. Both of these processes accelerated about 400 million years ago, yielding atmospheric oxygen comparable to its present abundance. Since that time the concentration has been essentially steady, as evidenced by fossil charcoal, indicating that it would support the combustion of wood. With the massive flux of oxygen through the atmosphere, the balance of production and loss appears to be delicate and such a prolonged balance implies a feedback mechanism, with the loss of oxygen directly controlled by its abundance. The oxygen production by photolysis would have been steady, satisfying a base load of oxygen demand, possibly at too low a level for advanced life forms, and any feedback control of the supply would have depended on the response of plant life to the oxygen abundance. But there is another possible explanation: the rates of consumption of oxygen by the weathering of crustal rocks and the decomposition of vegetation depend on oxygen availability, and offer just the sort of feedback mechanism that is required. They slow down when oxygen is less abundant.

Our interpretation of the oxygen story becomes more persuasive when examined quantitatively. We have mentioned that the oxygen demand by weathering of crustal rocks could account for an amount of oxygen equal to that in the present atmosphere in as little as 15 million years. This rate is characteristic of the last 500 million years or so of abundant oxygen. Before that, the oxygen demand would not have been fully satisfied and weathering would simply have consumed what was available. In the absence of vegetation protecting the land areas, the demand would have been higher and kept the free oxygen to a very low level. In spite of that, the cessation of banded ironstone formation 1.8 billion years ago indicates continuous availability of some free oxygen from that time. If weathering was consuming it at the same rate as in the last 500 million years, then the total consumed since then would have been 120 times the oxygen in the present atmosphere. Over the life of the Earth it would have amounted to 300 times, but this is more than can be explained by the production of oxygen by any plausible combination of hydrogen loss and photosynthesis. But we must consider a total of about 100 times the present atmospheric oxygen. The total production by photosynthesis, calculated in Chapter 12, on the basis of the abundance of buried organic carbon, would have provided 41 times the present atmospheric oxygen. As estimated in this chapter, hydrogen loss inferred from the deuterium/protium ratio would have yielded at least 59 times the present atmospheric oxygen. So, we conclude that the photolysis of atmospheric water vapour, with the loss of hydrogen to space, has provided a 'base load' of oxygen production for the entire life of the Earth, but for most of that time it was completely consumed by weathering. Photosynthesis introduced another supply and the combination is essential to our present oxygen abundance. One alone would not have sufficed.

Ozone

There is another possible link between the simultaneous appearances of abundant atmospheric oxygen and life on land and that is ozone. Fossil evidence of marine life predates the earliest land life probably by 2 billion years and, by some estimates by more than 3 billion years, although that is disputed [7.13]. The late onset of land life requires an explanation that is more convincing than the supposition that it simply awaited evolutionary development. It appears possible that the solar ultraviolet radiation that is intercepted by stratospheric ozone would have been too destructive of land life before there was sufficient atmospheric oxygen to produce the ozone. Almost all atmospheric oxygen is in the form of O_2, molecules comprised of two atoms, but, as discussed in Chapter 4, in the upper atmosphere some of these molecules are disrupted, releasing separate oxygen atoms that attach to other molecules, forming O_3. This is ozone, which has a crucial role in preventing a dangerous range of solar ultraviolet radiation reaching the surface. It appears possible that the sea provided sufficient protection for marine life before the appearance of stratospheric ozone [7.14]. We have argued that photolysis of water vapour has always produced some oxygen in the upper atmosphere, but when the concentration was extremely low, freed oxygen atoms would have had a very low probability of meeting up with an oxygen molecule to attach to. They would be more likely to attach to nitrogen and mix into the rest of the atmosphere, reaching the surface and oxidising crustal rocks in that form. The ozone layer depends on abundant oxygen for its existence and would not have been available more than about 500 million years ago, leaving open the possibility that early life on land developed as soon as it was protected by ozone from short wavelength ultraviolet radiation. That happened only when the available oxygen sufficed for production of the ozone.

Atmospheric argon

Another atmospheric gas that presents a clue to the evolution of the Earth and its comparison with Venus and Mars is argon, Ar. It is a chemically inert gas with three naturally occurring isotopes, of masses 36, 38 and 40 atomic mass units. All of them were incorporated in the Sun and planets when the solar system formed, but ^{40}Ar has increased subsequently as a decay product of the rare potassium isotope ^{40}K. ^{40}Ar is the isotope of immediate interest. ^{36}Ar/^{38}Ar $\approx$ 5.3 is, within the uncertainties of observation, the same everywhere in the solar system and we do not need to give ^{38}Ar separate consideration, but ^{36}Ar and ^{40}Ar show wide variations. Using the numbers in Table 7.1, we can isolate the abundances of ^{40}Ar in the atmospheres of the three planets. Expressed as parts per billion of total planet mass, as for the numbers in the table, they are 3.3 (Venus), 11.25 (Earth) and 5.8 (Mars). On the available evidence, especially from the Russian Venus probes, the abundances of potassium in the three planets are similar so the numbers reflect the effectiveness of argon leakage to the atmospheres. Planetary outgassing is essentially a volcanic process. Most of the argon released by volcanism appears in the emitted gases and its inertness allows it to diffuse rapidly from lava, ash, any hot rock, or its weathered and eroding remains. Its greater abundance in the Earth's atmosphere means that the tectonic processes on the Earth have been more effective in bringing it to the surface. Perhaps surprisingly, they have been more effective on Mars than on Venus and this prompts the remark that the difference is probably due to water. Mars has more evidence of past water, but the Earth has vastly more than either of them. Once again, we see that water, in this case as the lubricator of plate tectonics, has many roles as a vital ingredient of the environment.

It remains to comment on the remarkably high abundance of ^{36}Ar, the light isotope of argon, on Venus. Absorption from the solar wind, in which ^{36}Ar dominates, has been suggested, but there are unexplained variations in the other inert gases, neon, krypton and xenon. There is also difficulty with helium, which is abundant in the Venus atmosphere and in the solar wind. It is light enough to escape, but remains surprisingly abundant, whereas hydrogen, which is dominant in the solar wind, has almost completely disappeared. We leave these as unsolved problems.

8. Thermal balance, the greenhouse effect and sea level

Surface temperature is controlled by radiation

Although the Earth is a complex dynamic system of interacting parts, we can start developing an understanding of the thermal condition of its surface layers, atmosphere, oceans and crust, by first supposing it to be a solid body with its surface in thermal balance. Then the real Earth can be studied in terms of departures from this simple model. In this context, 'thermal balance' means primarily a radiative balance in which the Earth radiates back into space the energy that it receives from the Sun, with a negligible adjustment for heat from the interior. Any imbalance causes heating or cooling. This simple model gives a surprisingly good estimate of the average surface temperature of the Earth. It involves the concept of thermal radiation — that is, the radiation that a body emits by virtue of its temperature, discussed in Chapter 3 and especially Note [3.1]. The essential point about thermal radiation is that the total emission is proportional to the fourth power of absolute temperature (T^4, in Kelvin, degrees centigrade plus 273) making the process very sensitive to temperature. A doubled absolute temperature means a 16-fold increase in radiation. In Table 8.1, surface temperatures calculated by this model [8.1] are compared with observed planetary temperatures. We see that the model works well enough for the Earth and Mars to provide a starting point for our discussion, but is wildly wrong for Venus, and we examine the reason for that presently.

Table 8.1. Planetary surface temperatures.

Planet	Venus	Earth	Mars
Mean surface temperature	730 K (457°C)	288 K (15°C)	215 K (−58°C)
Mean black body temperature	327 K (54°C)	278 K (5°C)	225 K (−48°C)
Peak wavelength of thermal radiation*	4.0 μm	10 μm	13.5 μm

*Sunlight peak 0.5μm (micro-meters; millionths of a metre). Visible spectrum 0.43μm to 0.69μm. Ranges of absorption by CO_2, H_2O and O_3 are shown in Fig. 8.1.

Radiation is not all the same

The wavelengths of thermal radiation vary systematically with the temperatures of radiating bodies. Although all wavelengths are radiated, they have a spectrum with peak emission at a wavelength that varies inversely with the temperature. A hot body radiates at all wavelengths more strongly than a cooler one, but peaks at a shorter wavelength. The peak wavelengths for planetary temperatures are given in Table 8.1. They are all much longer than the peak of solar radiation because the planets are much cooler than the Sun. Our eyes are sensitive to the solar range, with a long wavelength limit of about 0.69μm, at the red end of the visible spectrum. In Chapter 4 we refer to solar ultraviolet radiation, which is beyond the blue/violet end of the visible spectrum. Most of this is absorbed in the upper atmosphere, where it causes ionisation, breaking atoms into electrically charged components (ions) and separating the hydrogen from oxygen in water molecules. The planetary thermal radiation is referred to as infrared (more red than red), having longer wavelengths, and is invisible. The atmosphere is almost completely transparent in the visible range, except where radiation is intercepted by clouds or aerosol haze, but is more opaque in both ultraviolet and infrared ranges. The infrared opacity arising from absorption bands of carbon dioxide and water, at wavelengths indicated in Fig. 8.1, results in the phenomenon that we call the greenhouse effect.

As seen in Fig. 8.1a, the range of wavelengths that we refer to as visible light is the range of strongest solar radiation. Evolution has made us sensitive to the radiation that is most useful to us. Many animals have adaptions that differ from ours in minor ways. Birds have sight extending into the near ultraviolet range, which aids the discrimination of fine details at a distance. At least some fish are sensitive to the polarisation of light (which we can identify only with optical instruments), and so distinguish refractions and reflections at the water surface. A few species of snake have infrared sensors by which they can identify the thermal radiation from warm-blooded prey. All such developments are driven by their evolutionary advantages, emphasising the point that biological evolution is controlled by the environment and, therefore, that changes to it that are too rapid can be disastrous.

Principle of the greenhouse effect

We think of the greenhouse effect as a modification of the radiative balance, with heating of the surface by incoming solar radiation, to which the atmosphere is transparent, and blocking of outgoing thermal radiation by atmospheric opacity. Although this is basically valid, it is simplistic. The reality is more complicated and the analogy to a greenhouse is imperfect. The glass of a greenhouse presents a barrier to air movement, retaining heated air, and, although its infrared opacity adds to the effect by blocking outgoing thermal radiation, that is of secondary significance. Temperature is controlled by vents at the top, allowing a restricted convective exchange with the cooler air outside. There is no comparable restriction on motion of the atmosphere, in which the buoyancy caused by temperature differences drives convection, with upward transport of heat. An important component of this is the latent heat of water vapour, which is released by condensation higher in the atmosphere. To understand how the greenhouse effect operates in this situation, it is convenient to refer to the extreme conditions of Venus, which has a massive CO_2 atmosphere (Table 7.1) and a very high surface temperature (Table 8.1).

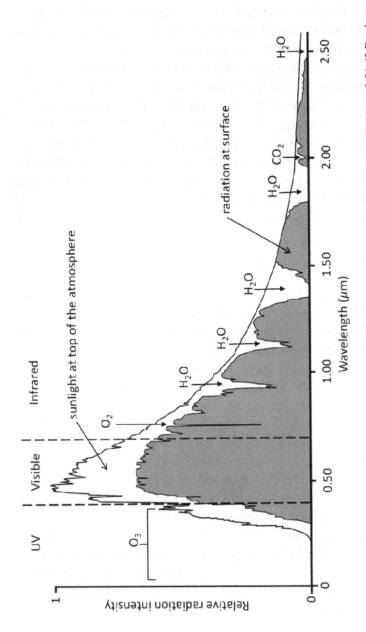

Fig. 8.1. Absorption of radiation by the atmosphere as a function of wavelength for (a) Incoming solar radiation and (b) (following page) Outgoing thermal radiation. The wavelengths absorbed by certain gases are marked. On a clear day approximately 70% of the solar radiation reaches the surface but only about 30% of the infrared (thermal) radiation from the surface escapes directly to space. The greenhouse effect is caused by the difference.

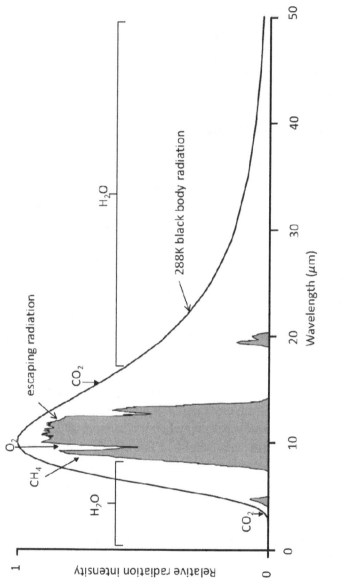

Fig. 8.1. (b)

Is the high temperature of Venus correctly described as a greenhouse effect?

We define the greenhouse effect as radiative heating of the surface with blocking of outgoing heat (whether radiative or convective). Although the Venus situation is commonly attributed to an extreme greenhouse effect, that seriously compromises the definition. The atmosphere of Venus is almost opaque to incoming solar radiation as well as outgoing thermal radiation. Its sulphurous clouds reflect sunlight very strongly and only about 5% reaches the surface as diffuse light from scattering in the dense cloud. It is also true that there is no evidence of 730 K thermal radiation, corresponding to the surface temperature, radiated from the planet, so some greenhouse action presumably occurs, even with the very weak sunlight. However, most of the solar radiation that is not reflected is absorbed high in the atmosphere and re-emitted as thermal radiation, but that does not reach the surface. It maintains a radiative thermal balance at a high altitude, roughly identified with the cloud tops, and the relationship of the temperature there to that at the surface is a property of the atmosphere and its circulation. We are familiar with the decrease in temperature with height in the atmosphere and that bicycle pumps heat up when the air in them is compressed. The air in the atmosphere is compressed by the weight of air above it, so that its pressure and density increase towards the surface. With vertical circulation, rising air expands and falling air is compressed. Unless the circulation is extremely slow, the heat transfer to or from any sample of air as it moves is negligible, so it cools as it expands and warms up when it is compressed, establishing a decrease in temperature with height in the atmosphere. So it is in the Venus atmosphere, which is moved about by vigorous winds. The relationship between the temperatures at the cloud tops and at the surface is a consequence of compressional heating. The resulting temperature gradient is referred to as the lapse rate. On Venus it is complicated by the fact that the lower part of the atmosphere is above the critical pressure of CO_2 (7.4 MPa), at which there is no distinction between liquid and gas. Also condensation and re-evaporation of atmospheric sulphuric acid reduces the lapse rate, in the same way as water vapour does in the

Earth's atmosphere. Presumably weak solar heating provides some convective driving force for the atmospheric circulation, but most of that would originate higher in the atmosphere where more energy is absorbed. In any case the circulation would very effectively convey the surface heat up to the cloud-top level of radiative balance and there is no way that the surface temperature could depart dramatically from an adiabatic relationship with the cloud tops. The black body model is simple and approximate, so it appears surprising that it gives temperatures that are as close to the observations as they are for Mars and more particularly for the Earth with its complex cloudy atmosphere. But both the black body model and the greenhouse effect are irrelevant to the surface temperature of Venus.

Greenhouse warming is controlled by water vapour

Our point that the near coincidence of the Earth's average surface temperature with the black body model is fortuitous is prompted by the fact that clouds screen much of the Earth from direct solar radiation. They also inhibit the outward transfer of thermal radiation, but the overall effect is cooling, so that a greenhouse effect is essential to the compensation for cloudiness. This occurs by a feedback mechanism involving water vapour [8.2]. Of the solar radiation that reaches the surface, some is simply reflected and most of that escapes because it meets the same atmospheric transparency on the way out as on the way in. The rest is absorbed by the surface, which emits thermal radiation, but most of that is absorbed by the infrared opacity of CO_2 and water vapour in the lower part of the atmosphere. The absorbing gases re-radiate the heat in all directions, returning some of it to the surface and establishing a radiative exchange, which results in an increase in temperature. At high temperature the air can hold more water vapour, so there is a positive feedback, reinforcing the temperature rise. The air is heated and expands, tending to rise convectively. How far it rises depends on its temperature contrast relative to the surrounding air, but we suppose, for the moment, that the surrounding air is cool enough to allow it to rise indefinitely. As it does so, it cools and eventually reaches a level at which its content of water vapour starts to exceed the saturated vapour pressure, the maximum that the air can hold at that temperature (Fig. 8.2). It starts to condense, forming cloud. The saturated vapour pressure depends very strongly on temperature and the cold upper reaches of the atmosphere can hold very little. Cloud formation is part of the process that recycles water and heat in the lowest 10 km of the atmosphere, the troposphere. The height at which cloud starts to form in the rising air depends on its starting conditions at the surface, temperature and relative humidity, the water vapour pressure expressed as a percentage of saturation (Fig. 8.2).

Although the condensation of water vapour, producing droplets (or ice crystals) in the rising air, causes an increase in density, because the water droplets or ice are denser than the vapour they replace, that is more than compensated by the thermal expansion of the air, caused by the release of latent heat by condensation. The air can continue to rise, building cloud to greater height with an increasing water or ice content and decreasing vapour. The cooling continues, but the decrease in temperature with height is reduced by the latent heat release. Whereas, until the saturation point is reached, temperature decreases by about 9.8°/km, which is referred to as the dry adiabat, above that point the temperature gradient is reduced to the 'wet adiabat', about 6.5°/km. Although temperature and humidity gradients vary widely, it is this wet adiabat that exerts the strongest control on atmospheric structure. Its significance is seen by reconsidering the rise of our (still dry) air, cooling at 9.8°/km. If the temperature of the surrounding air decreases more slowly with height, in particular if it follows the wet adiabat, then the rising air loses its buoyancy where the temperatures become equal and it ceases to rise, whether or not it has reached the height of the cloud base represented in Fig. 8.3. The establishment of a wide scale dry adiabatic temperature profile is prevented by the ubiquity of water vapour and the global average temperature gradient is close to 6.5°/km. This is a measure of the control exerted by water vapour on the structure and dynamics of the atmosphere.

With this interpretation of the thermal structure of the atmosphere, we have a basis for considering the overall energy balance

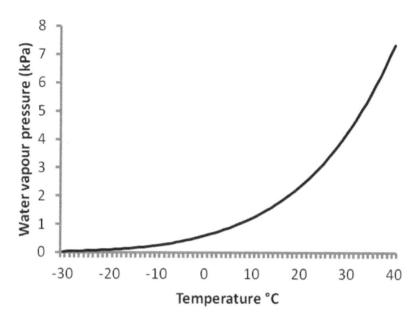

Fig. 8.2. The saturated vapour pressure of water, the maximum that air can hold, is very sensitive to temperature. The actual water content of any volume of air is a fraction of this, generally referred to as relative humidity, and expressed as a percentage. The total atmospheric pressure is 101.3 kPa.

and the greenhouse effect. As we have mentioned, thermal radiation from the heated surface does not get very far. The heat is transferred back to the surface and to the lowermost atmosphere, causing convection, but it is not usually dry convection. Surface water is evaporated and condenses at higher elevations, returning as rain (or snow), having released its latent heat higher in the atmosphere. The water vapour contributes to the buoyancy of the wet air because the water molecule (molecular weight 18) is lighter than other air molecules (average molecular weight 28.97). This mode of upward heat transport prevails in the troposphere, roughly up to 10 km altitude. At that level the water vapour is greatly reduced and the atmosphere is transparent enough for a radiative balance of thermal energy to take over. Now we see the similarity to the Venus

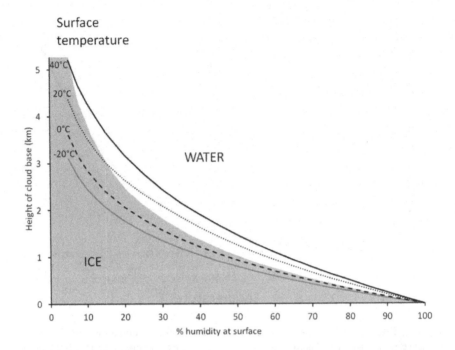

Fig. 8.3. The height of the cloud base, at which water vapour starts to condense out of convectively rising air, as dependent on its ground level conditions, relative humidity (% of saturation) and temperature. The curves for widely different temperatures are very similar, illustrating an essential role of relative humidity in atmospheric behaviour.

situation, with a radiative balance high in the atmosphere and atmospheric properties controlling the relationship to surface temperature. But there is a crucial difference. Solar radiation reaches the surface of the Earth and thermal re-radiation of the energy is inhibited by the atmosphere, a true greenhouse effect, but very little solar radiation gets near to the surface of Venus.

Up to this point we have concentrated on the role of the water vapour feedback in the greenhouse effect and need to emphasise that it is just that — a feedback mechanism. Increased atmospheric water vapour cannot be identified as a cause of global warming, but as an effect of warming. It is a condensing gas, with an atmospheric content controlled by temperature. If there is some other cause of a temperature rise, such as an increase in solar radiation or in atmospheric opacity, caused by non-condensing greenhouse gases (carbon dioxide, methane), then atmospheric water vapour will also increase and will amplify the change. But the increased water vapour is a response to warming initiated by some other effect. This feedback must be seen as an effect, not a cause, of the warming. In identifying the cause of current global warming, we must concentrate on CO_2, with the effect of methane as secondary. In the lower atmosphere CO_2 is less effective than water vapour (Fig. 8.1), but both are important greenhouse gases. In the stratosphere and higher, where there is little water vapour, it has a greatly diminished effect, but CO_2 is as large a fraction of the atmosphere as lower down and takes over, but it shares the role of greenhouse gas with ozone, which is concentrated in the stratosphere.

A negative greenhouse effect in the stratosphere

The combination of CO_2 and ozone in the stratosphere has an interesting effect that provides convincing evidence of its significance. Whereas the lower atmosphere is warming, the stratosphere is cooling. This is a negative greenhouse effect to which both CO_2 and ozone contribute, although both have positive effects in the lower atmosphere. There the incoming solar radiation warms the surface, which emits thermal (infrared) radiation that is intercepted by the atmospheric opacity and repeatedly re-radiated within the atmosphere, retaining the heat. But the low density and dryness of the stratosphere give it a lower infrared opacity and some of the upgoing radiant heat escapes to space, resulting in cooling. The greenhouse effect of the infrared opacity of both CO_2 and ozone is positive, causing warming, in the lower atmosphere, but negative in the stratosphere, which is cooled. Figure 8.4 presents an interesting illustration of the combined effects of ozone and CO_2 in the stratosphere. In the period of reliable (satellite) determinations of the temperature of the lower stratosphere, encompassing the height at which ozone is most abundant, there were two major explosive volcanic eruptions, which injected a mix of gases (and ash) into the stratosphere. An important volcanic product is sulphur dioxide (SO_2), which oxidises and forms droplets of sulphuric acid that reflect sunlight and cause

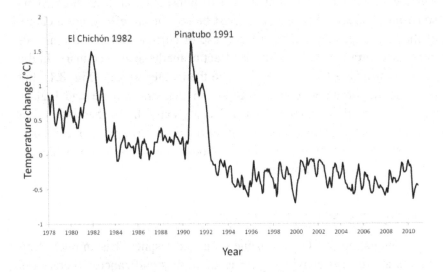

Fig. 8.4. Satellite observations of temperature variations in the lower stratosphere, January 1979 to October 2011. This is a global average (82.5°N to 82.5°S) of data that show some latitude variation, but also more scatter for restricted latitude ranges. For each of the two major volcanic eruptions that occurred in this period, the initial effect is stratospheric warming by destruction of some of the ozone that causes greenhouse cooling, but a longer term effect is cooling by additional CO_2. Data from NOAA Satellite and Information Service, collected by Remote Sensing Systems (RSS).

cooling of the surface and troposphere, as discussed further in Chapter 9. But in the more tenuous stratosphere, the volcanic materials destroy ozone in the same way as the chloro-fluoro-carbons (CFCs) that were widely used as refrigerants and spray propellants until embargoed by international agreement. When stratospheric ozone is reduced in this way, there is less greenhouse cooling, resulting in the positive peaks in Fig. 8.4, which endured for about two years, until the inhibiting chemicals were washed out. The volcanic CO_2 that reached the stratosphere was more long lasting, so that, when the ozone returned to 'normal', there was still an enhanced CO_2 contribution to stratospheric cooling and not a simple return to the pre-eruption temperature. The opposite effects in the stratosphere and troposphere are persuasive evidence of the greenhouse effect as the cause of tropospheric warming because any increase in insolation, by increased solar luminosity or orbital effects (Milankovitch cycles, Chapter 3), would cause warming of both the troposphere and stratosphere, which is not what is observed.

The approach to quantitative calculations of the greenhouse effect

Ideas about greenhouse-mediated global warming have a long and tortured history [8.3], beginning in 1824 with papers by the French mathematician-physicist, Joseph Fourier. He argued that the Earth is warmer than would be expected from a simple balance of incoming and outgoing radiation and that this could be explained by a greater opacity of the atmosphere to the outgoing thermal radiation than to the solar input. His argument was based on observations of the rate of cooling of hot bodies and was more than 50 years ahead of the first presentation of the relevant radiation equations [3.1], but was notable for its recognition of the physical principles. It remained an unconfirmed hypothesis until 1862, when the British physicist, John Tyndall, measured the opacities of various gases to thermal radiation. He found that dry air was transparent to the radiation, but that water vapour, coal gas (of which methane is a major constituent) and carbon dioxide all absorbed it. The next important step was in 1896, when Swedish scientist Svante Arrhenius made some detailed numerical calculations (revised in 1901) on the effect of a changing atmospheric content of CO_2, including the feedback effect of water vapour. Although the data used by Arrhenius were rough and incomplete, his estimate of a surface temperature rise of 4°C to 6°C by doubling of the atmospheric CO_2 content is within the range of modern estimates, the principles of which are discussed in Chapter 15. Even at the end of the 19th century, one of Arrhenius's colleagues, Arvid

Högbom, estimated that the production of CO_2 by human activity was comparable to the amount recycled by some natural processes, but this was not regarded seriously at the time because interest in this work was in trying to understand ice ages and not in human influences on climate. It was not until 1938 that a British engineer, Stewart Callendar, boldly declared that the burning of fossil fuels was measurably warming the Earth. But war and other troubles intervened and when interest in the problem resumed, it was met not just by doubt and indifference but by vigorous, and sometimes dishonest, attempts to discredit it, and these are ongoing [8.4]. The seriousness of Callendar's warning was eventually given a grudging political nod when, in 1988, the Intergovernmental Panel on Climate Change (IPCC) was formed, nominally by the World Meteorological Organisation, a UN agency. With a mix of scientific and political input, the panel's voluminous reports have been predictably cautious, recognising the reality of greenhouse warming, but generally being, perhaps, too careful to avoid overstating its consequences.

Clouds moderate the greenhouse effect

We mentioned cloudiness and need to give it further thought. The reflection of sunlight back to space from cloud tops has a bigger effect on global temperature than the interception and re-radiation back downwards of thermal radiation from the Earth, giving cloudiness a general cooling effect. Would rising temperatures cause their own negative feedback by increasing evaporation and consequent cloud cover, moderating the temperature rise? The obvious answer is yes, but different kinds of cloud have different effects and satellite data on total cloud cover have not distinguished them adequately to prove the point conclusively. However, it is clear that the role of clouds is central to the atmospheric cycles of water and heat and needs close attention. In discussing the effect of cloudiness on radiation in the atmosphere, we need a wider definition than the extent of visibly obvious clouds. Widely

Joseph Fourier, 1768–1830

French mathematician and physicist best known for the mathematical technique of Fourier analysis, by which a repeating series of numbers can be resolved into a sum of sinusoidal waves (Fourier series). He was politically active and had an extended military career with Napoleon. He established the basic equation of heat conduction and was the first, by many years, to recognise the principle of the atmospheric greenhouse effect. A bust by Pierre-Alphonse Fessard in Grenoble, France.

John Tyndall, 1820–1893

British physicist, who, as director of the Royal Institution in London (a position made famous by Michael Faraday and the discovery of electromagnetic induction, the basis of the electrical industry), published numerous books making the principles of physics accessible to a wide audience. Following early experiments on magnetism, his most important contributions were to the study of radiant heat, in the course of which he measured infrared absorptions by gases that are now central to concern about greenhouse warming.

Svante Arrhenius, 1859–1927

Swedish physical chemist with pioneering contributions to the understanding of ionic compounds (such as salt, NaCl), and to the energies of molecular reactions. With delayed acceptance of his early work, he turned to some astrophysical ideas, including the cause of ice ages. Using optical properties of carbon dioxide and water vapour, in 1896 he made the first detailed calculation of greenhouse warming of the Earth (revised in 1901), producing what is still an acceptably accurate estimate.

distributed aerosols impart some opacity where there are no distinct clouds and they are having an increasing effect, dubbed global dimming. The significance of the contribution by human activity to atmospheric aerosols is evident from the fact that the northern hemisphere is affected more than the southern hemisphere. An observation that makes it appear comparable in importance to visible clouds is the well observed progressive decrease in the daily temperature range (DTR), that is, the difference between daily maximum and minimum temperatures [8.5]. This is a consequence of the faster rise of minimum temperatures than maximum temperatures and is less variable, and so more clearly observed, than variations in the mean temperature. The increasing aerosol-induced opacity of the atmosphere restricts the diurnal range by reducing both incoming and outgoing radiation.

Cloudiness is subject to its own feedback control. Evaporation of water depends on the temperature, humidity and wind speed of adjacent air, but is increased by direct exposure to sunlight, which ejects molecules from a water surface. Since increased evaporation leads to increased cloud, reducing the sunlight, this is a negative feedback which, on a global (but not local) scale, stabilises the cloud cover. Aerosol dimming of the atmosphere biases this effect, reducing evaporation and hence global rainfall.

Ocean warming and sea level rise

Global warming means increasing temperatures of surface and near surface features, the uppermost crust, lower atmosphere and oceans, all of which accumulate and store added heat. The total amount of this heat is a measure of the departure of the Earth from thermal equilibrium. At the present time, by far the largest proportion (85% to 90%) is stored in the oceans and even the heat applied to the melting of glacial ice is included in the other 10% or 15%. Heating and thermal expansion of the oceans is one of the causes of sea level rise. The present total rate of the rise, about 3.2 mm per year (Fig. 8.5), has several causes, of which the most obvious is melting of glaciers and grounded ice caps. At the present time it is the north polar cap that is melting fastest, but that is floating and has no significant effect on sea level [8.6]. A detailed study of the causes of sea level rise over the period 1961 to 2008 [8.7], which included allowances for the opposite effects of pumping ground water and the storage of water in dams, concluded that thermal expansion of sea water resulting from rising ocean temperatures contributed 0.8 mm/year to the sea level rise, with a net heat input to the oceans of 165 terawatts (5.2×10^{21} J/year). We use the numbers from that analysis to assess the long term significance of the departure from thermal balance implied by rising ocean temperature. Other studies have given slightly different numbers, but we can draw some very obvious conclusions without precise numbers.

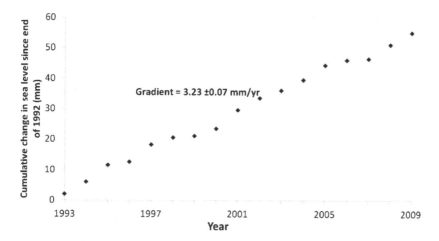

Fig. 8.5. The global average rise in sea level, as measured by satellite altimetry, timed radar reflections to satellites from the sea surface. The data points are annual mean values and the uncertainty in the linear gradient is 1 standard deviation. Since satellite orbits are controlled by the Earth's gravity, these measurements determine sea level relative to the Earth's centre of mass. Longer term records from tide gauges are less secure, but indicate that the present rate of rise is steeper than in earlier decades. Data by TOPEX/Poseidon and Jason 1 project teams (www.podaac.jpl.nasa.gov).

First, we point out that the importance of ocean warming to sea level rise is intuitively surprising. Discounting the human-induced effects included in the study of Note [8.7], and considering only the contributions of ocean warming and ice melting in that study, 53% of the sea level rise is attributed to ocean warming and 47% to ice melting. Since the peak of the last ice age, sea level has risen by about 120 metres, and if we imagine that 53% of that is also to be attributed to ocean warming then it requires thermal expansion to have caused a 1.65% increase in the total ocean volume [8.8]. That is off the scale of Fig. 8.6. If the entire ocean was at maximum density at the peak of the ice age, then, for that much thermal expansion, it would now be at 60°C throughout. But, except for shallow water in tropical and low latitudes, we still have a cold ocean (Fig. 6.9). No more than a trivial fraction of the 120 m rise of the sea can be attributed to its thermal expansion. The overwhelmingly dominant contribution was ice cap melting. Similarly, thermal expansion made a negligible contribution to the sea level rise after the previous glacial maximum [8.9]. The distribution of heat at the current stage of global warming is fundamentally quite different from what happened during recovery from ice age glaciations. Most of the greenhouse-induced heat is absorbed by the oceans, but this is a transient effect, peculiar to the present physical conditions of the Earth and the current greenhouse warming mechanism, for a reason that we discuss in Chapter 15. In the long term contemplation of a major sea level rise, the concern must be ice cap melting and not ocean thermal expansion.

We have another perspective on the ocean expansion question by relating the expansion to the heat input. Taking the volume increase per year, $\Delta V = 2.9 \times 10^{11}$ m^3 (0.8 mm rise over 3.62×10^{14} m^2 of sea), to be caused by the quoted annual heat input $\Delta Q = 5.2 \times 10^{21}$ J, the ratio, $\Delta V/\Delta Q = 5.57 \times 10^{-11}$ m^3/J, requires the water that is heated to have an average thermal expansion coefficient of 2.3×10^{-4} K^{-1} [8.10], which, from Fig. 8.6, implies an average water temperature of 23°C. Only a very small fraction of the ocean is at or above this temperature (Fig. 6.9), so these numbers require that virtually all of the quoted heat input goes into the shallow tropical ocean. While this may be consistent with the very modest heat input required for the ice melting contribution to sea level rise, it is an extreme conclusion and suggests that either the heat input to the oceans has been underestimated or its contribution to the sea level rise has been overestimated (and the melting contribution underestimated). The inferred heat distribution is emphasised by considering the sea level rise that would occur if the ocean heat input (5.2×10^{21} J/year) were, instead, applied to ice melting. The answer is

45 mm/year [8.11]. Per unit of heat input, ice melting is far more effective in causing sea level rise than thermal expansion (by the factor 57 with these numbers). But, approximate as it may be, the estimated ocean heat input provides a quantitative measure of the effect of greenhouse warming.

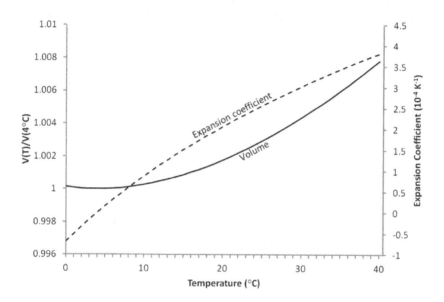

Fig. 8.6. Thermal expansion of water. The solid curve shows the variation in volume with temperature, relative to the volume at 4°C, the temperature of maximum density, and the broken curve gives the expansion coefficient. Sea water data differ little from these fresh water values, but with maximum density at 2°C.

Ocean warming is a transient effect

As itemised in the analysis cited in Note [8.7], the net input of solar energy to the Earth is about 0.2% of the energy that reaches the surface (Table 8.2). This is a non-trivial departure from thermal balance and cannot be maintained for long without very dramatic consequences. The magnitude of this heat input is obscured to casual observation by the very large heat capacity of the oceans (emphasised also in Chapter 7). But, as we have pointed out, this distribution of the heat input is transient. It is a consequence of the departure from equilibrium of the temperatures of the atmosphere and ocean. The ocean monopolises the heat input because its large heat capacity makes it slow to respond and it has not caught up with the rising atmospheric temperature. How long would it take to catch up if the greenhouse-induced heat stabilised? This is a question to which we can give only a very approximate answer, based on the atmospheric and oceanic heat capacities and the present heat partitioning: 130 years by an estimate in Note [15.5]. But the concept of a catch-up is not immediately relevant as long as greenhouse heating continues to increase. We have to ask 'Are we heading towards a new equilibrium state and, if so, how different will that state be from the conditions we are familiar with?'

Table 8.2. Surface processes: some energy comparisons. All values in terawatts. See also Table 14.1

Solar energy, top of atmosphere	174 000
at surface	98 000
Net input of solar energy	
to atmosphere (with	7
water vapour)	
to oceans	180
to ice melting	
grounded	9
floating	9
to continents	4
total	209
Heat from the interior	46
Polewards transport of heat by ocean currents	~2000

In a new equilibrium state there would be no continuing heat input to the oceans or atmosphere and the atmosphere would be at a higher temperature, radiating away all the received energy. Atmospheric modellers address the question 'How much higher?' We have mentioned that the first serious attempt to answer it was an 1896 calculation by Arrhenius, who concluded that atmospheric CO_2 doubling would cause a 4°C to 6°C temperature rise. This is at the high end of the range of modern estimates, but it set a standard for the comparison of alternative models, which have adopted CO_2 doubling as a reference benchmark. A very simple model that illustrates the principles, discussed in Chapter 15, gives 2.6°, at the low end of the range. But we are not now in an equilibrium state and the atmospheric temperature rise is moderated by the oceans. In an equilibrium state, with no changes in the mean temperatures of either the atmosphere or oceans, the atmospheric temperature will suffice to convey all received heat to the upper atmosphere by more vigorous convection, with increased evaporation and rainfall. This will entail an upward movement of the upper boundary of the moisture-controlled lower atmosphere (the tropopause), where a new radiation balance will be established. At this point in the argument, clouds become critical and we need to learn more about them. But we can make a general observation that they have a negative feedback effect and the temperature rise will be accompanied, and moderated, by increased cloudiness. This moderation will be essential to the control of climate change.

Interpreting ice cap contraction

Returning to the role of the oceans in slowing the temperature rise, as equilibrium is approached and they stop absorbing the net greenhouse-induced heat, it will be redistributed in two ways. One is simply an increase in outgoing radiation in response to the higher temperature, and the other is melting of the ice caps. In Chapter 15 we point out that the absorption of heat by the oceans actively inhibits the transfer of heat to the polar caps, especially in the southern hemisphere. The melting of polar ice is recognised as one of the indicators of global warming and it will dramatically accelerate as the ocean-atmosphere temperature balance is approached. The present rate must be regarded as anomalously

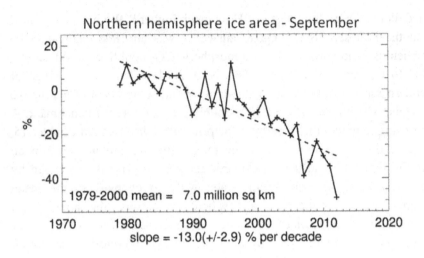

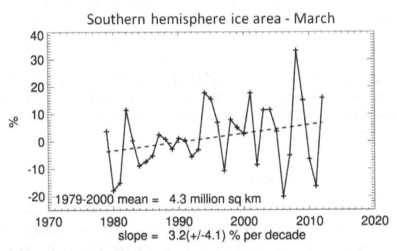

Fig. 8.7. Percentage changes in areas of sea ice in the polar regions at times of maximum sea ice retreat in successive years, relative to the averages for 1979–2000. Data from National Snow and Ice Data Center (NSDIC.ORG).

low. It is fastest for the north polar cap, which is not just floating but drifting and its 'export' of ice into the North Atlantic contributes to its shrinking. Figure 8.7 contrasts the decrease in the sea ice of the north and south polar caps. Since sea ice is floating ice, it includes all of the north polar cap but only a small fraction of the Antarctic ice, most of which is grounded. But the opposite trends in the figure require an explanation because, at first sight, they suggest that global warming is restricted to the northern hemisphere. That is not so. Southern hemisphere glaciers are retreating in the same way as those in the north, and some melting of land based Antarctic ice is needed to explain the total sea level rise. The north polar ice is identified as either multi-year ice or first year ice. Sometimes some of the ice is classified as 'seasonal', but this is not strictly the same as 'first year' because some of the first year ice survives a summer melt and becomes multi-year ice. First year ice has increased in the Arctic and is now the dominant form. The decrease in total arctic ice mass is attributed to a major decrease in multi-year ice [8.12]. The fact that first year ice is increasing with the decrease in multi-year ice in the north appears consistent with the increasing sea ice and ablating grounded ice in Antarctica, where the area of sea ice is increasing although its total volume is decreasing, as first year ice expands but older and thicker ice diminishes (as in the Arctic). Global warming is global and sea level rise is an inevitable consequence. While there is a strong temperature imbalance between the atmosphere and ocean, the rate of the rise will remain low, but that is just a feature of the transient state we are living in. The delay in the onset of a much more rapid rise will continue while there is accelerating greenhouse warming, because of the 'protection' of the ice by the ocean heat capacity, and will be at least 100 years, but the longer the delay the more dramatic the rise will be.

9. Environmental crises and mass extinctions of species

Discontinuities in the fossil record

The geological periods listed in Table 9.1 were established by the sequence of fossil-bearing sedimentary rocks long before radiometric dating added reliable ages. Each of the named periods is identified with particular groups of fossil species and the breaks between them mark sudden changes. In general, within each period there are progressive changes, consistent with gradual evolution, but the breaks mark discontinuities, with many disappearances, followed by proliferation and diversification of survivors. The geological periods identify intervals of relative quiescence with no dramatic changes and, by inference, stable environments. The breaks between them indicate environmental crises, when evolution took a step back and looked for new opportunities. Species that may have been very successful in their stable environments, but could not adjust to sharply changed conditions, disappeared, presenting to survivors an opportunity for expansion and more rapid evolution to occupy new or freed up environmental niches. The causes of the environmental crises have been debated for as long as the evolutionary discontinuities have been recognised. The most nearly complete extinction event in the 300+ million year period of well developed fossils occurred 251 million years ago at the boundary between the Paleozoic and Mesozoic eras (Table 9.1), when more than 90% of all species disappeared, with dramatic reductions in the numbers of the surviving species, leaving the sedimentary record almost barren for several million years. The next most dramatic event occurred 65 million years ago, when about 2/3 of all species disappeared. This event, at the Cretaceous-Tertiary (K-T) boundary, has attracted most attention, partly because it marked the end of the reign of the dinosaurs, but also because it became a focus of debate over the causes of mass extinctions.

Table 9.1. The geologic time scale 2004 [9.1] Numbers are dates of commencements of geological periods (millions of years ago). The most dramatic discontinuities coincide with eruptions, at the times marked with asterisks, of flood basalts in the Deccan area of India and in Siberia.

Phanerozoic	Cenozoic	Quaternary	Holocene	0.01	
			Pleistocene	1.81	
		Tertiary	Pliocene	5.33	
			Miocene	23.03	
			Oligocene	33.9	
			Eocene	55.5	
			Paleocene	65.5	*
	Mesozoic	Cretaceous		145.5	
		Jurassic		199.6	
		Triassic		251	*
	Paleozoic	Permian		299	
		Carboniferous		359	
		Devonian		416	
		Silurian		444	
		Ordovician		488	
		Cambrian		542	
Proterozoic				2500	
Archean				4000	
Priscoan (Hadean)					

The asteroidal impact hypothesis

The debate became heated in 1980, when L. W. Alvarez and collaborators published evidence of a massive asteroidal impact coinciding with the K-T boundary [9.2]. Their original evidence was a layer of clay in an exposed section of the boundary at Gubbio, in Italy, in which they found an anomalous abundance of iridium. This is a common minor constituent of meteoritic iron. Subsequent work has left little doubt that that there was a major impact at that time, and even identified the impact site at Chicxulub on the Yucatan Peninsula of the Mexican east coast. There followed a strenuously argued claim that the impact caused the K-T extinction event, presuming that, in throwing debris and volatiles around the world, it darkened the sky sufficiently and for long enough to cause dramatic global cooling. This provoked a sometimes bitter debate with proponents of a rival explanation: extinction events are caused by volcanism.

The volcanic interpretation

The K-T boundary also coincided with a great outpouring of basaltic lava in what is now the Deccan area of India. A rapid series of massive flows, many metres thick and extending for hundreds of kilometres, accumulated to a layer at least 2 km thick in less than half a million years, 65 million years ago. Erosion has exposed the basalt layers as a series of steps and they are referred to as 'trapps', from a Scandinavian word for steps, but often written as 'traps'. Similar formations occur in other places, notably the Parana trapps in Brazil and, most impressively, the Siberian trapps. In each case the lava appeared in a geologically short time and, at least in most cases, the time coincided with one of the boundaries in Table 9.1. The coincidences between the dates of the various trapps and the boundaries in the table have been documented by Vincent Courtillot [9.3] and are summarised in Fig. 9.1. In view of the inference that flood basalts caused the extinctions, we need to look closely at the nature and cause of flood basalts to see how they present a viable mechanism for mass extinctions.

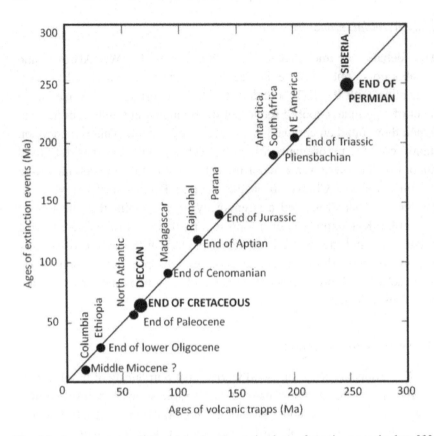

Fig. 9.1. A comparison of the dates of major extinctions of species over the last 300 million years with the ages of massive flood basalts on land, referred to as 'trapps'. The two most dramatic extinction events, coinciding with the two most impressive of the 'large igneous provinces', are distinguished by bold lettering. After Courtillot [9.3].

There are basically three types of volcanic activity, illustrated cartoon-style in Fig. 9.2. Two of them occur in the plate tectonic process, discussed in Chapter 5. This provides the cooling mechanism for the mantle and drives most geological activity, producing the crusts of the ocean floors and continents, with representative compositions in Table 1.2.

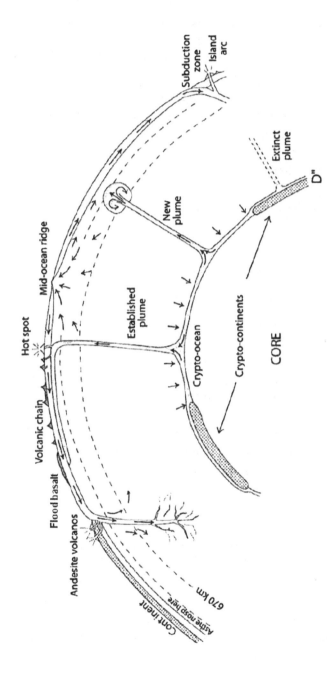

Fig. 9.2. A cartoon of mantle convection, showing three stages in the life of the deep mantle plumes that convey core heat to the surface.

In Chapter 5 we refer also to the isolated 'hot spot' volcanoes, of which Hawaii is the most studied example, that are not parts of the plate tectonic cycle. They are the surface expressions of deep plumes that carry heat up from the core, at a depth of nearly 2900 km. It is these plumes that are identified with flood basalts. Like the ocean ridge basalts they are primary volcanic products of the mantle, with similar silica contents, essentially different from the secondary products that comprise the continents. The violently explosive volcanoes occur along the subduction zones, where infusion of sea water introduces a substantial volatile content to the lava. Plume basalt is less viscous and erupts more smoothly but, as the flood basalts demonstrate, it can erupt in much larger volumes.

Origin and effects of flood basalts

The core is about 1000° hotter than the mantle above it and produces a hot, softened layer of rock at the bottom of the mantle (layer D", dee double primed, in the jargon of geophysics). This is the source of plume material. Being hot, it is much less viscous than the mantle material above it and, being buoyant, it flows upwards through narrow channels, referred to as plumes, or plume stems. As it rises, decompression causes partial melting, producing the basaltic lava that erupts from the hot spot volcanoes. However, the D" layer is not a steady source of plume basalt. It is not only mobile, it is heterogeneous and probably includes 'rafts' of dense material, floating on the core and identified as 'crypto continents' in Fig. 9.2. As the D" material flows towards the plume bases, it carries the rafts with it and, when one moves directly under a plume it reduces, or even cuts off, the plume flow. The starved plume may die, although slowly, and then the D" buoyancy initiates a new one, but it does not have a prepared channel and must force its way up through the mantle with a plume head several hundred kilometres in diameter, fed by a plume stem of fresh hot material. The rise of a new plume is necessarily slow, taking at least 20 million years for the plume head to reach the surface, but when it does so it includes partial melt amounting to a few million cubic kilometres of basaltic lava, ready to erupt. The lava flows that become flood basalts drain the plume head, but the flow up the plume stem does not stop. As the surface plate, now carrying an extensive layer of flood basalt, drifts away, the plume remains more or less stationary and repeatedly punctures the crust, producing a series of volcanoes, of which the Hawaiian Islands are the classic example. In many cases the chain of volcanoes can be traced back to the initiating flood basalts [9.3]. Of particular interest are the flood basalts identified with the Cretaceous-Tertiary extinction event 65 million years ago that now form the Deccan plateau in India, which lead, via the Maldive Islands, to the currently active plume marked by the volcano on Reunion Island, in the Indian Ocean.

Sulphur dioxide as the cause of volcanic cooling

As mentioned in Chapter 8, there are numerous reports of global cooling following volcanic eruptions [e.g. 9.4]. Particularly well documented is the 1991 eruption of Mt Pinatubo, in the Philippines, which ejected dust and gas with a high concentration of sulphur dioxide (SO_2) into the stratosphere. Mixing with water vapour and oxygen, it is converted to a haze of sulphuric acid, reflecting sunlight. If the SO_2 does not rise past the lower atmosphere it is fairly quickly washed out as acid rain, but in the stratosphere it can survive for a few years and is the most significant agent of volcanically induced cooling. Although the ash from violent eruptions, such as Pinatubo, has commonly been blamed, SO_2 survives much longer in the stratosphere and is now considered to be more significant. A particularly large explosive eruption of this kind occurred about 73 000 years ago, leaving a massive crater at Toba, in northern Sumatra, and causing widespread environmental stress — but it is dwarfed in scale by the flood basalts. Although basaltic eruptions are more passive than the acid volcanism of the subduction zones, they produce SO_2 which becomes widespread in the atmosphere, as evidenced by the large basalt flows of the 1783 Laki (Iceland) eruption, from which SO_2 has been identified in Greenland ice cores. An anomalously cold European summer that year was attributed to the Icelandic volcanism by Benjamin Franklin, then American ambassador in Paris.

It is not necessary for an eruption to be explosive to put SO_2 into the stratosphere. The largest of the great flood basalt flows had volumes of at least 10,000 cubic kilometres, extending for many hundred kilometres in all directions, and approaching 1000 times the largest of the Laki flows. An atmospheric thermal plume from such a large hot area would certainly have carried fumes well into the stratosphere and caused deep, prolonged, SO_2-induced cooling. A further effect would have arisen from the emission of CO_2, which has a more permanent atmospheric residence, and would have caused greenhouse warming when the SO_2 eventually washed out. The eruptions would have been separated by extended warm periods, perhaps lasting for hundreds of years, between the repeated episodes of sudden cooling. The sizes of the eruptions and the abundance of SO_2 in basaltic lava leave no doubt that the cooling events were dramatic and the coincidences with times of mass extinctions make a direct causal connection appear obvious.

We can take a closer look at the atmospheric effects of very large volcanic eruptions to see why the plume basalts would have been so effective at causing environmental crises. Injection of SO_2 into the stratosphere is crucial and the well documented case of the Pinatubo eruption has given some quantitative evidence. The stratospheric haze was perceptible for as long as 5 years after the eruption, but decreased with time in the manner of an exponential decay, which means that it had a characteristic decay time, referred to as a half-life, in this case a period of 9 months. The effect was reduced by a factor of 2 after 9 months, a factor of 4 after 18 months, and so on. Explosive subduction zone eruptions such as that of Pinatubo, and presumably of Toba, the biggest documented case, produce sudden large injections of SO_2 into the stratosphere, causing cooling that lasts for at least a year. But such high concentrations of SO_2 produce an acid haze with larger droplets than lower concentrations. Large droplets fall out faster than small ones and they also reflect sunlight less effectively. Their cooling effect is both weaker and shorter-lived than for similar volumes of small droplets. By contrast, the massive plume basalt flows of the Deccan or Siberia would have occurred over extended periods, at least months and probably years, and would have been fuming for many more years, producing stratospheric haze of fine droplets over long periods. We have referred to the environmental effects of the Laki eruption of 1783, which produced a series of flows similar to those of the great flood basalts, but many times smaller. Scaling up the Laki flows by a factor 100 or 1000 leaves no doubt that the environmental consequences of flood basalts suffice to explain the mass extinctions.

Flood basalts will occur again

Although the case for volcanism as the cause of mass extinctions appears unassailable, can we admit asteroidal impact as another possible cause? If so, the case for it rests on the impact coinciding with the K-T boundary and the volcanic explanation suffices for that anyway [9.5]. A major impact could have had a global effect lasting for as long as a year, but was the K-T extinction event as sudden as that? Opinion is divided, but impacts coinciding with other extinctions are hard to find. If iridium layers are one of the indicators, then there is negative evidence in several iridium layers with no corresponding extinctions. But it is not necessary to categorically rule out the impact hypothesis. It can be left 'on the table', while recognising the much greater strength of the volcanic evidence. This has a bearing on the danger of further, repeat performances. More impacts will occur, but the supposition that they will cause mass extinctions relies on the less than convincing evidence that they have done so in the past. On the other hand, the core continues to lose heat, as it must to maintain the Earth's magnetic field (Chapter 4), and that means that new mantle plumes are inevitable. It should be possible to see them coming with more warning than an asteroid would give, because plume heads take millions of years to traverse the mantle. Could seismology distinguish them from other mantle heterogeneities? Possibly. The best candidate meriting attention appears to be a volume of high temperature (identified by low seismic wave speeds) beneath the East Africa rift valley.

Benjamin Franklin, 1706–1790
American statesman with very wide interests and
activities, including printing and publishing. The best
known of his numerous scientific investigations was a
demonstration (at some personal danger) that lightning
is an electrical discharge. As an inventor, he was
responsible for the diving bell, used for work under water.
He was the American ambassador in Paris from 1776 to
1785, spanning the time of the major volcanic eruption of
Laki (Iceland) in 1783. He identified that as the cause of
persistent haze, an early end of summer, with extensive
crop failures, and an unusually cold following winter in
Europe. Portrait by Joseph-Siffred Duplessis.

The punctuation of evolution by environmental crises

The mass extinctions present a commentary on evolution, as seen in
paleontological studies of the development of fossilised creatures. The
process of natural selection, or evolutionary competition, often referred
to as 'survival of the fittest', operated differently during periods of
environmental quiescence and crisis. The concept of 'fittest' requires
somewhat different definitions in these two situations. Especially during
the late stages of well established geological periods, the rules of
competition between members of a species or between predator and prey
remained essentially constant. Gradual evolutionary changes were driven
by the advantages of size, speed, etc., and creatures with specialised
needs could flourish. But these attributes are not what is needed to
survive an environmental crisis. Then the need to withstand, or hide
from, harsh conditions, with a minimum demand for particular food, may
make smallness an advantage. Adaptability becomes crucial and
specialised requirements are a hazard. The ability to take advantage of
new situations ensures not just survival but prosperity in a changed world
with reduced competition. Much has been made of the disappearance of

the dinosaurs at the K-T boundary, but they did not really disappear completely. The specialised monsters vanished, leaving us with birds. As the most adaptable of the dinosaur line of descent, birds have not merely survived but proliferated.

10. Stability of the environment

Identifying feedback mechanisms that control the environment

The Earth has supported life in some form for at least 2 billion years (up to 3.8 billion years by some estimates) and advanced forms for the last half billion years. Life has evolved as the Earth, atmosphere and oceans evolved, but the fact that it exists at all imposes some basic requirements that have been continuously satisfied for much of the 4.5 billion year existence of the Earth. The limited range of surface temperature is a crucial factor and we give that particular attention. The later developments that make the Earth liveable for advanced life forms, including humans, are partly a consequence of life itself. This is an example of a phenomenon that is a central theme of this chapter: feedback. This is any process by which an effect reacts on and modifies its cause. We can regard the fact that primitive life forms paved the way for more advanced forms as a case of positive feedback — life facilitating its own development. The environment is subject to a variety of feedback mechanisms, both positive and negative. In general, stability results from negative feedback, the tendency to counteract any imposed change. Positive feedback reinforces a change and may lead to instability. Although environmental stability relies on a dominance of negative feedback mechanisms, there are some very obvious positive ones and we consider how they are counteracted. We discuss first a few natural positive feedbacks that appear to have the potential to be environmentally calamitous, but from which the Earth has recovered, although generally on a very long time scale. Can we learn enough from them to be appropriately cautious about activities with environmental effects that are subject to positive feedback processes, now that human activity has global scale geophysical consequences?

Glacial cycles and temperature stability

Ice ages have come and gone and will almost certainly do so again (unless prevented by very strong greenhouse conditions). For the last million years, periods of extensive ice cover (ice ages) have lasted for thousands of years, with the appearance of being cyclic in Fig. 13.2, but there were earlier periods with glacial conditions lasting for millions of years; the most recent of these were the Permian period, 300–350 million years ago, and the late Precambrian, shortly before 550 million years ago (also giving the vague impression of being cyclic). The vastly different time scales between the recent and earlier ice ages indicate different causes, but the effect is basically the same and it invokes a much discussed positive feedback. With large areas of both hemispheres ice-covered and reflecting sunlight, solar heating is reduced, reinforcing the cooling that initiated the ice ages. How did the Earth recover? There is no secure answer to that, but the easiest suggestion to understand is that ice ages are both initiated and terminated by variations in the output of the Sun or orbital effects (Milankovitch cycles), with terrestrial effects such as greenhouse conditions and cloudiness as secondary, feedback consequences. If that is the correct interpretation then, at the termination of an ice age either the Sun must not just return to 'normal', but must over-compensate to defeat the positive feedback of reflective ice cover, or else glacial conditions must bring on their own negative feedback, perhaps by an effect on the atmosphere.

Compensation for the faint early Sun

As we have mentioned, the starting point for a stable environment is a more or less steady global average temperature. Even ice ages cannot be regarded as extreme departures from this condition, bearing in mind that there is a temperature range exceeding 50°C over the Earth at any time.

The evidence of simple biological life billions of years ago means that there were liquid oceans at that time. We argue in Chapter 1 that zircons with ages exceeding 4 billion years, perhaps even up to 4.4 billion years, required a liquid ocean almost as soon as the Earth had a stable structure. Although it may not then have been suitable for any form of life, it indicates a temperature not differing greatly from present conditions. But, as discussed in Chapter 3, models of the Sun and its nuclear processes give solar energy increasing with time, as its core contracts by replacement of hydrogen with denser helium, resulting in a present output about 30% greater than 4 billion years ago. The difficulty in rationalising this conclusion with the evidence that there has always been a liquid ocean constitutes the faint early Sun paradox. There appears to be little prospect of a significant revision of the solar calculations, so resolution of the paradox requires one of three things: a progressive increase in the radius of the Earth's orbit, another energy source that has decreased with time, or atmospheric conditions very different from those at present, perhaps with a very strong greenhouse effect. We examine each of these but conclude that the first two must be discounted and that some way must be found to stabilise the temperature with changes to the atmosphere.

Stability of the Earth's orbit

In addition to the loss of mass by the Sun, which we discount as insignificant in Chapter 3, there are three effects that cause the radius of the Earth's orbit about the Sun to increase, (i) solar tidal friction, (ii) drag of the solar wind (the magnetic centrifuge mechanism) and (iii) asymmetric re-radiation of received solar energy (the Yarkovsky effect). To compensate for a 30% change in solar luminosity, the distance to the Sun would need to change by about 15%, that is, 22 million kilometres. We examine the orbital radius changes resulting from each of these three effects to see whether they could, in principle, cause a change in orbital radius as big as that.

(i) The lunar tide is bigger than the solar tide (Fig. 2.4) and its energy dissipation has caused a major change to the orbit of the Moon (Fig. 2.3). The solar tide dissipates only about 20% as much energy at present, and an even smaller fraction in the distant past when the Moon was closer. It causes the Earth's orbit about the Sun to expand, in the same way as the Moon's orbit about the Earth expands. (In each case, it is more correct to say that the two bodies are each in orbit about their shared centre of mass. Reference to the orbit of one about the other is an approximation used when the masses are very different, but the tidal torque operates on both). However, the angular momentum of the Earth's rotation is much less than a millionth of its orbital angular momentum [10.1], and no conversion to orbital angular momentum can have a material effect on the orbit. The increase in orbital radius in the life of the Earth caused by friction of the solar tide is less than 100 km. The effect of the tide raised in the Sun by the Earth is even less significant.

(ii) Although the Sun has more than 99.8% of the mass of the solar system (assuming that we know about it all), its rotation accounts for little more than 0.5% of the total angular momentum. In suggesting an explanation for this, the Swedish astrophysicist, H. Alfvén, noted that young solar-type stars have very strong and extensive magnetic fields, with vigorous stellar winds, and that this must have been the case also with the youthful Sun. The magnetic field would have rotated with the Sun, dragging with it the surrounding nebula of accumulating planetary material that would still have been highly ionised by strongly radioactive debris of a then recent supernova, making it an electrical conductor. We refer to this as the magnetic centrifuge mechanism. Although the Sun's magnetic field is now much weaker, it still exists, and there must be a continuing centrifuge effect, so the question arises: does its drag on the Earth suffice to cause a significant expansion of the orbit? The Earth is not an ionised particle but a solid body with its own magnetic field, so the force exerted by the solar field takes effect as a force between magnetic fields. But it is weak and is causing the Earth's orbit to expand by only about 3 km per billion years. Details of the calculation are given in Note [10.2].

(iii) Solar radiation also exerts a force, termed radiation pressure, on all bodies in the solar system. It is directed outwards from the Sun and so diminishes slightly the effect of gravitational attraction to the Sun. This is an important effect for small particles, with cross sectional areas (intercepting the radiation) that are large relative to their masses, but not for larger bodies, and in any case a purely radial force is not of great interest in the present context. However, there are secondary effects arising from the absorption and re-radiation, or reflection, of the solar radiation. One of these, termed the Yarkovsky effect after its discoverer, does influence the orbits of larger bodies [10.3]. A rotating body absorbs solar radiation on its sunlit side and emits the energy as thermal radiation with a delay, cooling as it does so. This means that it radiates more strongly from the recently heated 'afternoon' hemisphere than from the cooled 'morning' hemisphere, and there is an imbalance in the momentum carried away by the thermal radiation, exerting a net force on the 'afternoon' hemisphere. If the body is rotating in the same sense as its orbital motion, as in the case of the Earth, that force increases its momentum and causes the orbit to expand. There is no simple theory that can deal with this problem rigorously for a body like the Earth, but we can make a simplifying assumption that greatly exaggerates the magnitude of the effect to put an extreme upper limit on what is possible. If we assume that the Earth absorbs all of the sunlight falling on it and re-radiates the energy only after it has rotated by 90°, then the orbital expansion would be 31,000 km per billion years [10.4]. Noting that, even by this gross overestimate of the effect, it fails by a wide margin to come near to resolving the faint early Sun paradox, we abandon attempts to explain away the paradox by orbital evolution.

Internal heat is of no consequence to the surface temperature

The heat flux from the Earth's interior is gradually decreasing in the manner that would be required to offset the increasing solar radiation, but it is far too little to have a significant influence on the heat balance at the surface. The present global heat flux is 44 or 46 terawatts and we estimate that it would have been 5 times as much 4 billion years ago. But that is still insignificant compared with the 174,000 terawatts of solar energy at the top of the atmosphere, or about 98,000 terawatts reaching the Earth's surface. This is obvious from the numbers in Table 8.2. Translating these numbers to the local scale, vertically incident bright sunlight delivers more than a kilowatt per square metre to the surface, but heat from the interior averages only 65 milliwatts per square metre over land areas. Although the internal heat is responsible for the tectonic effects that contribute vitally to the environment, the heat itself has no observable surface consequences outside the limited areas of current volcanism. This point is emphasised by the fact that more internal heat is released on the ocean floors (106 mW/m^2 on average) than on continents, but the ocean floors are kept cold by circulation of water from polar regions. It is difficult to discern climatic effects of the variation in solar power over the 11-year sunspot cycle, about 0.15% or 260 terawatts, but this variation is much greater than the flux of internal heat (although comparable to the heat flux 4 billion years ago). Thus, we are driven to seek compensation for the weak early Sun by atmospheric effects, because of the lack of alternatives, not because of direct evidence.

The faint early Sun paradox is more serious for Mars

We may wonder just how serious the paradox really is, a point we refer to presently, but first note that it is more serious for Mars than for the Earth. Erosion features on Mars point to the existence of flowing water in the distant past, clearly impossible at the present average surface temperature, about −58°C. These features indicate a hydrological cycle, with water vapour, rain and oceans as the source of the water. Mars has cooled dramatically, against the trend of increasing solar output, whereas for the Earth we seek only a compensation for the solar trend. In appealing to some variant of the greenhouse effect, the case of Mars compels us to recognise that a CO$_2$-rich atmosphere, which Mars has, does not suffice. Something else is needed. In Table 8.1 we see that Mars has a surface temperature slightly lower than suggested by the simple black body radiative balance [3.1], in spite of an atmosphere completely dominated by CO_2. Expressed as mass of CO_2 per unit area, the Martian atmosphere has 1676 kg m^{-2}, compared with Earth's 6.1 kg m^{-2}. Its behaviour is similar to that of the Earth's stratosphere, which is also tenuous, cold and dry. As mentioned in the greenhouse discussion in Chapter 8, in this situation CO_2 may cause cooling, not warming. As a wet planet, with abundant surface water as well as its CO_2-rich atmosphere, Mars could have had a strong greenhouse effect, but not strong enough to compensate for both the weak early Sun and the fact that it receives less sunlight than the Earth by a factor 2.3. Another possibility is that Mars had an opaque atmosphere, similar to that on Venus.

Plausibility of a strong early greenhouse effect

A greenhouse explanation for the temperature of the early Earth can be invoked, without the need to postulate Venus-style atmospheric opacity. If we adjust the numbers in Table 8.1 for a solar luminosity that is 70% of the present value, following the black body radiative law summarised in Note [3.1], the mean black body temperature of the Earth is reduced from 278 K to 255 K (−18°C) [10.5]. The evidence for an early ocean does not require it to have been as extensive as the present oceans. A tropical ocean would have sufficed and the global average temperature may well have been as low as during any of the ice ages. A temperature 10° higher than the black body value would have allowed substantial liquid oceans and that is within the range of a greenhouse explanation. It requires more atmospheric CO_2 than at present, but we know that the Earth had that as well as water. This argument can be extended by pointing out that the Earth would now be very cold without the atmospheric CO_2 and that the early Earth would probably have been in the snowball state even with the present atmospheric CO_2. Geological evidence points to a high CO_2 concentration in the early atmosphere, as well as abundant surface water. This means that, in attributing the early surface temperature to greenhouse warming, we have a confirmation that, given a 'wet' Earth, temperature rises with increasing atmospheric CO_2. It is a useful confirmation that increasing atmospheric CO_2 causes warming of a 'wet' Earth.

The ice core record: evidence of a natural CO_2 balance

Having attributed the current global warming to CO_2, we need to consider the natural controls on its abundance. The correlation between CO_2 and temperature is particularly well-documented for the last few hundred thousand years, as displayed in Fig. 13.2, but the cause-effect relationship is not clear and, at times, appears to be reversed, in the sense that temperature changes led the CO_2 variations. What is more significant is that the peaks in CO_2 concentration were all very close to the pre-industrial average value, 280 ppm. We conclude that, within very close limits, the total CO_2 content of the atmosphere and oceans remained constant over the period of the ice age record in Fig. 13.2. The variation in atmospheric content resulted from the temperature dependence of the solubility of CO_2 in sea water. Cold sea water dissolves more CO_2, reducing the atmospheric concentration and in Chapter 13 we point out that the relationship between CO_2 and temperature is quantitatively explained in this way [13.4]. The changes were more rapid than possible variations in the ocean content of calcium that could have sequestered CO_2 as carbonate, so it is evident that the sources and sinks of CO_2 external to the atmosphere-ocean system were balanced. That leaves the apparently cyclic glaciations seen in Fig. 13.2 in need of an explanation independent of the greenhouse effect, and the approximate 100,000-year repeat time is strongly suggestive of orbital variations, the Milankovitch cycles, mentioned in Chapter 3. These include a 96,000-year oscillation in the ellipticity of the orbit (its departure from a perfectly circular shape), although it is not clear that this is a sufficient explanation of the temperature changes.

For the duration of the record available from ice cores, volcanic emissions of CO_2 were within the range accommodated by the natural feedback and there was evidently no dramatic influx, such as must have accompanied the extinction events discussed in Chapter 9. The sources of CO_2 have been balanced by the sinks, the most important of which is the formation of carbonate rock in the oceans. It relies on the influx of calcium dissolved from the continents and by hydrothermal circulation at ocean ridges, net of the calcium subducted with ocean crust, as considered in Chapter 6, which we estimate to amount to about 200 million tonnes per year. This suffices to fix the CO_2 from natural sources, assuming volcanic emissions at the geologically recent rate (and allowing for submarine volcanism), but is less than 1% of what would be needed to neutralise the CO_2 from the present rate of fossil fuel burning. That is upsetting the balance in a way that cannot be accommodated by natural processes, being well outside the range of the negative feedbacks that control the CO_2 abundance. There is an important implication in this, which we draw to attention again in Chapter 15. The CO_2 that we release will remain in the atmosphere-ocean system indefinitely.

Stability of the atmospheric oxygen concentration

In Chapter 12 we estimate that the total mass of buried organic carbon in the Earth, derived by photosynthesis from atmospheric carbon dioxide, corresponds to the release of free oxygen amounting to about 41 times the present atmospheric oxygen. It follows that this carbon could burn only by consuming the same amount of oxygen. But most of it occurs widely dispersed, in very dilute forms. The mass of carbon in all of the accessible deposits of coal, oil, natural gas and some oil shales exceeds 10^{16} kg (10 thousand billion tonnes) and if all of this were burned it would consume nearly 3% of the atmospheric oxygen. Of course, it would also produce CO_2, and if that all remained in the atmosphere (ignoring solution in the oceans, which would diminish as ocean acidity increases), it would raise the content from the 2010 value, 390 ppm, to 4700 ppm. Unimaginable as this may be, it emphasises that our concern is with CO_2, which must remain a minor component of the atmosphere, and that the abundance of oxygen is not seriously endangered by this prospect. However, on the subject of oxygen, the Earth has had an oxygen-rich atmosphere only for about the last 10% or so of its existence. It is certainly unique in the solar system and requires special explanations (Chapter 7), making it appear fragile. Its approximately constant abundance for the last 400 million years requires control by a sensitive feedback process, also outlined in Chapter 7. There is a balance between production and consumption, controlled by the abundance, and the only obvious abundance-sensitive process is the oxidation of crustal rocks, primarily on land but also on the sea floor.

A return to the question of cloudiness

As we mentioned in Chapter 8, of the feedback mechanisms that have a stabilising effect on climate, the one that is least understood is probably cloudiness. A rising global temperature means increasing evaporation, and therefore also increasing precipitation because 'what goes up must come down'. Since precipitation requires clouds, they must change, either becoming more widespread, thicker, differently distributed in altitude or latitude, or all of these. Clouds are not all the same with respect to reflective properties, and cloud cover surveys (by satellite or otherwise) cannot readily discern their effectiveness in controlling the radiation balance. But a general increase as the Earth warms is inevitable and the negative feedback on global temperature changes is a consequence. It must apply to both greenhouse warming and varying luminosity of the Sun, although not necessarily equally effectively.

11. Inorganic mineral deposits as products of an evolving environment

Development of the continental crust is the first stage

The continental crust contains a wide diversity of minerals, some of them in remarkable concentrations. Although it has only 0.4% of the Earth's mass, it has more than half of the total Earth content of some elements — including potassium, which is of particular interest because of its radioactivity. Many of the minor elements, including the other two thermally important radioactive elements, uranium and thorium, are also very strongly represented in the crust, where they have been further concentrated into exploitable ores by processes that require very particular environmental conditions. A dramatic example is the uranium deposit at Oklo in Gabon, West Africa, where uranium leached from neighbouring rocks was washed into a series of ponds and accumulated in such concentrations that two billion years ago the ponds became natural nuclear reactors [11.1].

The abundant elements in the crust, as listed in Table 5.4, include two of the commercially interesting metals selected for inclusion in Table 11.1. This table lists the factors by which natural processes must concentrate these metals, relative to the average upper crust, to produce economically viable ores. Even in the case of aluminium, for which the factor is quite small because this is the most abundant metal in the crust, the necessary concentration requires very particular environmental conditions, discussed below. Some of the others are products of sequences of processes which required different environments in turn.

Table 11.1. Factors by which metals must be concentrated by natural processes, relative to their average concentrations in the upper continental crust, to produce exploitable ores.

Aluminium	4
Iron	9
Copper	90
Zinc	130
Lead	1200
Gold	3000
Uranium	11000
Mercury	100 000

The appearance of the continental crust was the essential first stage in the development of many mineral deposits and, as pointed out in previous chapters, that depended on the oceans. We point out in this chapter that, in most cases, the subsequent processing to produce exploitable mineral concentrations also relied on abundant surface water. Since it is evident that the oceans have existed for more than 4 billion years, this is the time scale over which mineral deposits have been developing. Seen on this time scale, the present rate of their exploitation will be very short-lived. This chapter offers a perspective on that situation in the context of the diverse environmental situations that have produced them.

What is an inorganic mineral deposit?

There is an unavoidable arbitrariness in selecting mineral types that can be described as inorganic. Biological life has had a role in the development of the environment for billions of years, most intensively for the last several hundred million years, making it difficult to discern surface processes and effects that are not influenced by it, at least to some extent. But we see a need to distinguish the biological products that have affected the carbon-oxygen balance. Thus, limestone formed from the shells of marine creatures is clearly a biological product. It combines carbon dioxide with calcium dissolved in the sea and has had a massive impact by removing an enormous quantity of carbon dioxide from the atmosphere. But the carbon and oxygen are not separated in this process and it does not satisfy our very restrictive definition of an organic mineral deposit. This is reserved for materials, discussed in the next chapter, such as coal and oil, with reduced organic carbon, that is, carbon that has been separated from the oxygen in carbon dioxide. They have special environmental implications that are not shared by the minerals considered in this chapter. There are many types of mineral deposit that meet our inorganic definition and we select three very different ones to illustrate the range of environmental controls on their formation.

Hydrothermal activity

In Chapter 7 we discuss the development of banded iron formations by exposure of ferrous iron dissolved in the ocean to infusions of oxygen. The starting point for that story is the appearance of ferrous iron in the oceans in the first place. Sea water percolates through cracks in the ocean floor at, and near to, the ocean ridge spreading centres, where fresh igneous crust is forming. Contact with the rock raises the temperature of the water (as high as 500°C) and acidifies it, so that it becomes corrosive and dissolves minerals from the rock, including iron. Even with the weight of dissolved minerals, the buoyancy of the hot water suffices to drive it upwards, back into the cold ocean, causing some of the minerals to precipitate out as sediment on the nearby ocean floor, with some forming chimneys of mineral-rich rock, while cold water takes its place in a continuous cycle. Under the present conditions of an ocean with dissolved oxygen, these minerals include iron, but when the banded ironstones were forming there was little or no oxygen in the water and the iron remained in solution, spreading out over wide areas of the ocean to await the arrival of pulses of oxygenated water. The total length of spreading centres around the world's oceans is about 67,000 km, so ridge mineralisation is obviously widespread, although not conveniently accessible to exploitation. However, in a few places tectonic activity has exhumed former ridges and one of these, perhaps the most striking one, is on the island of Cyprus, in the eastern Mediterranean. Generally ocean ridges do not survive indefinitely because, being parts of the ocean floors, they disappear in subduction zones within about 100 million years of their formation. But the Mediterranean has had a complex tectonic history, alternating between a spreading centre and a zone of convergence, which it is now. The convergence has raised the sea floor, exposing ridge mineralisation at Mount Troodos on Cyprus, which is probably the best display of this anywhere, with related deposits at the nearby Mathiati mine.

Hydrothermal circulation occurs wherever water meets hot rock, and many, perhaps most, of the exploitable mineral deposits have hydrothermal origins or hydrothermal stages in their development. Volcanism is necessarily involved and the ocean ridges release a large fraction of the Earth's internal heat. Subduction zone volcanism is even more significant in terms of diversity of the mineral concentrations, but the principle is the same. As we have mentioned, volcanic gases and the weathering of crustal rocks consume atmospheric oxygen. This is a consequence of the fact that the deep Earth is in a low oxidation state. The hydrothermal plumes dissolve minerals in that state. There is normally also some sulphur present and, in the absence of available oxygen, the minerals are precipitated as sulphides, forming what are known as volcanogenic massive sulphide (VMS) deposits. These are sources of numerous metals, including copper, zinc, lead, gold, silver, molybdenum and tungsten. Although hydrothermal activity is an on-going geological process and these minerals are certainly forming at the present time, the process is so slow that renewal of such resources is not occurring on a human time scale. Their availability must be estimated in terms of what already exists. Iron presents a special case because, with the oxygenated ocean, no more massive banded iron formations will occur at all, but the known deposits are vast and this is not an imminent resource problem.

Placer deposits

In many situations the products of eroded rock are transported by stream flow, shoreline waves and wind, and the different densities and grain sizes lead to sorting and concentration of the various components. This is most significant in leading to accumulations of the denser minerals that collect in hollows in stream beds, slow flowing stretches of winding rivers and shoreline irregularities. Such accumulations are termed placer deposits and are valuable because they require little ore processing to extract minerals of interest. Placer deposits are important sources of tin, titanium, chromium, gold, platinum and even some gem stones, such as diamond, ruby and sapphire. Beach sands can be classified as placer deposits — particularly sand derived from eroded granite, which provides useful concentrations of rutile (titanium dioxide) and zircon (zirconium silicate) as well as nearly pure quartz. Most recognised placer deposits are geologically young, but there are important exceptions, notably the Witwatersrand gold deposit in South Africa, which appears to have accumulated in a reducing environment, before the appearance of free oxygen, at least 2.7 billion years ago.

The mention of diamonds in placer deposits raises an interesting point because diamonds form in only two ways, by deposition of carbon from vapour, which is irrelevant to the terrestrial situation, and by exposure of carbon to high pressure (and high temperature), corresponding to depths greater than 150 km in the Earth [11.2]. They are found in 'pipes' (volcanic conduits) of a basic igneous rock called kimberlite, which evidently erupted too quickly for the diamonds brought up with it to degrade to graphite on the way. Diamonds in placer deposits have been eroded from nearby kimberlite. The kimberlite deposits themselves are all small (normally less than a cubic kilometre) and are found only in the cratons or ancient cores of the continents. Although the kimberlitic intrusions are not as old as their surroundings, the diamonds in them appear to be so and their ultimate origin is an interesting matter for conjecture. In spite of being quite small, kimberlite eruptions were obviously vigorous and probably very brief events, initiated deep beneath the continents by accumulation of volatiles, especially water, in stable areas remote from plate boundaries where normal volcanism provides an escape route. Diamonds were found in placer deposits long before kimberlite pipes were recognised and their existence was enigmatic, because they could not have formed in or near the crust. They provide a good illustration of the role of water transport in the selection and exposure of minerals.

Laterite deposits

In old continental areas, worn down and now well removed from further rapid mechanical erosion, rocks may be exposed to very prolonged weathering, especially in the tropics with heavy rainfall and consistent high temperatures. These conditions give high biological productivity. Plant debris completely decomposes, generating acid soils; plant and tree roots, as well as burrowing animals, turn over the soil, exposing fresh rock to attack by fungi, microbes and the organic acids that they produce. With heavy rainfall, water movement removes soluble weathering products and even readily dislodged grains, leaving the minerals that are most resistant to weathering. Over tens of millions of years they form layers many metres thick that are referred to collectively as laterites, which may be derived from a wide range of rock types. They accumulate a select group of elements, particularly aluminium, iron and manganese. Greatest interest is in the layers dominated by alumina (aluminium oxide) and known as bauxite, because this is the only economic source of aluminium. In some usage the word laterite is reserved for the layers dominated by iron, but bauxite deposits are commonly coloured red by iron and may be capped by a layer of iron ore, so the distinction is blurred. Tropical soils are often enriched in aluminium, indicating that the leaching of more soluble components is quite general in the wet tropics. The time scale of laterite development is very long and may extend through several climatic changes that are reflected in the structures of the deposits. However, there are no bauxites with ages approaching the ages of the banded ironstones. The oldest known bauxites date from the Devonian period, which began 416 million years ago. Being surface deposits subject to erosion, it is reasonable to presume that any earlier bauxites have now disappeared. But the fact that their age range coincides with that of the land plants invites the supposition that they began to form more readily when rotting land vegetation acidified surface water.

Multistage mineral development requires diverse environmental conditions

Very few mineral deposits are products of single stage processes. They have appeared because the Earth is an active, evolving planet, providing diverse environmental conditions at different times and places. Just how diverse can be judged by visiting the mineral collections in museums and universities, but the larger scale, economic deposits are probably better indicators of global scale features of the environment and changes in it. We return to mentioning the banded ironstones, which present a particularly good example of multi-stage mineral production accompanying the evolution of the Earth. When the Earth was young there was effectively no free oxygen, and hydrothermal circulation accompanying sea floor volcanism put dissolved iron into the oceans. When oxygen became available the iron precipitated out. Tectonic activity subsequently raised the sea floor, exposing the ironstone on land.

Many useful minerals are sulphides, compounds with sulphur, which is an element that has chemical properties similar in many ways to oxygen. Sulphides form in the absence of available oxygen, as is commonly the case with hydrothermal deposits. In the presence of sufficient oxygen, oxides may form instead, with sulphur itself oxidising to sulphur dioxide, SO_2, and this occurs also when exposed sulphide ores are weathered. However, geochemists classify certain elements, including copper, zinc and mercury, as chalcophile, meaning 'sulphur-loving', and these are the ones most often found as sulphides. This draws attention to the fact that abundant oxygen is a surface, and near surface, phenomenon. The bulk of the Earth is in a reduced (oxygen-deprived) state and its exposure to the atmosphere, as fresh rock or volcanic gases, is continuously absorbing oxygen. Coupled with the observation that the oxygen-rich atmosphere is unique in the solar system, this reinforces the point that its origin requires very careful scrutiny (Chapter 7). Even at the surface, oxygen-free (anoxic) environments can develop, particularly in stagnant water in which decomposition of vegetation is halted by lack of oxygen. This is a subject of the next chapter. Another crucial observation, re-emphasised by the examples in this chapter, is that water is essential to almost everything that happens in the development of mineral deposits, starting with its role in plate tectonics and the formation of the continental crust. Another factor that may be less obvious, but could be important to the wide diversity of minerals in the crust, is the appearance of biological life, which has contributed to the formation of many minerals, not only those that are specifically recognised as organic and are the subject of the next chapter. The correlation of bauxite formation with land vegetation is an example.

12. Fossil fuels, buried carbon and photosynthetic oxygen

A comment on the carbon cycle

As mentioned in Chapter 7, the contribution of photosynthesis to atmospheric oxygen is difficult to judge with certainty. It requires an assessment of the total of the permanent burial of plant material that incorporates the carbon extracted from atmospheric CO_2. The large deposits of coal found on all continents demonstrate a contribution of land plants, but these do not come close to the required total. Marine plants and algae that fall to the sea floor, where they are carried to subduction zones and deeply buried, make another contribution to atmospheric oxygen, in which we include the oxygen dissolved in the oceans. However, for much of this material sequestration is temporary. As mentioned in Chapter 5, marine sedimentary material, subducted with sea water and as hydrated minerals, reappears in volcanoes within a few million years and CO_2 is one of the volcanic products; alternatively, if the buried carbon stays within the crust it is subject to the 75 million-year erosion-sedimentation cycle, also referred to in Chapter 5. It appears that no more than 10% of the buried carbon has a longer residence time in the mantle, and we can regard the subduction of organic carbon as part of a slow carbon cycle, with only a modest net contribution to atmospheric oxygen.

How much organic carbon remains sequestered?

This chapter addresses a question left hanging in Chapter 7: how necessary is it to find a source of atmospheric oxygen in addition to photosynthesis and the burial of carbon-rich organic material? The Earth's atmosphere is believed to have had, progressively if not all at the same time, a mass of CO_2 comparable to that now in the Venus atmosphere (Table 7.1). The disappearance of most of it is accounted for by the abundance of carbonate rocks (limestone etc.) in the crust. Absorption of atmospheric CO_2 in the formation of these rocks had no effect on the oxygen which remained bound to the carbon, so that the low present abundance of atmospheric CO_2 gives no direct information about the appearance of oxygen. That requires the separation of carbon from the oxygen in CO_2, to produce what we refer to as reduced or organic carbon. The end products of this process in the Earth are compounds of carbon with hydrogen, the hydrocarbons. As a general observation, the formation of hydrocarbons from buried plant material requires very specific conditions, sufficiently so to infer that only a small fraction develops into coal, oil or natural gas. Exploitable deposits are studied intensively, and so we have a reasonable idea of their total abundances. But in assessing the possibility that photosynthesis and the burial of fossil carbon suffice to maintain the atmospheric oxygen, by estimating the total reduced carbon, we must look past the economically interesting deposits, which are only a small fraction, and use the insight that has followed the study of the fossil fuels to consider the fate of buried organic material more generally.

The annual cycle of atmospheric CO_2 seen in the data in Fig. 13.1 shows that the rate at which it is absorbed by vegetation, in the northern hemisphere growing season, greatly exceeds the rate needed for maintenance of atmospheric oxygen. But most of the vegetation just decomposes within a year or so, consuming oxygen and releasing the carbon back into the atmosphere as CO_2. The essential question to be asked is: does the total store of reduced carbon that has been retained within the Earth suffice to explain not just the present atmospheric oxygen, but what has been absorbed by weathering of crustal rocks? The possibility rests on the abundance of widely distributed, dilute organic carbon that has no potential as exploitable fuel and so has not been assessed in detail by the exploration industry. Nevertheless, we can calculate how much fossil carbon is implied by the total oxygen release and express it in terms of an equivalent mass of high grade coal. If the present atmospheric oxygen (1.22×10^{18} kg) is to be explained as a residue of photosynthesis, and we imagine that the deposited carbon compounds have now degenerated to a layer of high grade coal in the crust, then the layer would be 0.9 m thick over the entire Earth, including the ocean floors (5.1×10^{14} m^2) [12.1]. The total that must be contemplated to produce all of the oxygen that has been consumed in weathering of the crust corresponds to a layer at least 50 m thick. Although this conveys a sense of the magnitude of the problem, it is simplistic, and we take a closer look at the evidence.

Photosynthesis releases oxygen from water as well as carbon dioxide

First, we note that the oxygen released by photosynthesis is not just a consequence of the extraction of carbon from CO_2; water is involved. The fossil fuels, as decay products of organic materials, are composed of more than carbon and, as mentioned above, the basic ingredients are hydrocarbons — compounds of carbon with hydrogen. The hydrogen has been extracted from water and that also releases oxygen. Water, and the release of oxygen from it, are central to the chemistry of photosynthesis and of plant material, but major changes to it occur during conversion to fossil fuels (or more widely distributed buried carbon) and our interest is in the end result: the ratio of hydrogen to carbon in the remaining sequestered hydrocarbons. This ratio varies from almost zero for the highest grade of coal to an atomic ratio of 4:1 for the lightest hydrocarbon, methane, CH_4, the major constituent of natural gas and a problematic component of coal seams. The net outcome of the production of a methane molecule from carbon dioxide (CO_2) and water (H_2O) is to combine the carbon from a molecule of CO_2 with the hydrogen from two molecules of water (H_2O), releasing two molecules of oxygen (O_2). In the notation of chemistry, this is made obvious by the equation

$$CO_2 + 2H_2O \rightarrow CH_4 + 2O_2$$

In this simple example, twice as much oxygen is released as would be inferred by the assumption that it is just the result of extracting the C from CO_2. The use of methane as a fuel reverses this equation, producing 60% as much CO_2 as burning coal to produce the same thermal energy [12.2]. This is a reason for favouring natural gas, rather than coal, as a fuel, and a motivation for seeking an environmentally acceptable method of extracting coal seam gas, about which we say more presently. We allow for this in making comparisons of the oxygen release implied by different carbon reservoirs, in terms of equivalent layers of coal.

Coal

This is generally recognised as the most abundant fossil fuel. It has resulted from the burial of land vegetation that has decomposed in low oxygen (anoxic) environments, such as deep bogs and stagnant lakes. The first stage is peat, much of which is identified as grasses, with larger plants and trees incorporated as they became available. Peat can be regarded as the lowest grade of coal. Carbon deposited as peat in the last 10,000 years amounts to 500 gigatonnes (5×10^{14} kg) globally. In developing into the higher grades of coal, peat loses most of the volatile material and is compacted by a factor of 5 to 10. The 10,000-year accumulation corresponds to a global layer of coal 1 mm thick. The recognised reserves of coal, of various grades, would amount only to a 7 mm thick global layer, although with the inclusion of deposits that are too deep, too small or of too low grade to be of commercial interest the layer would be more like 4 cm thick (compared with the 90 cm that would be required to explain the atmospheric oxygen as by-product of coal formation). There are two immediate conclusions from these numbers:

(i)   The carbon sequestered in the Earth as coal comes nowhere near what is needed to explain the atmospheric oxygen, and

(ii)   At the recent rate of peat accumulation, if it were all converted to coal the world's coal deposits would have developed in 400,000 years and not the 350 million years seen in Fig. 12.1(a). It is implausible that the rate of peat development in the last 10,000 years was so much faster than the average for the last few hundred million years; only a small fraction of peat develops into coal.

The temperature and pressure of burial convert coal into successively higher grades ('ranks' in American usage), with increasing proportions of carbon and rejection of the volatiles that contain hydrogen, especially methane. The lowest grade, lignite or brown coal, is only a step away from peat and has even been eaten by cattle. It is typically 40% or so water and only 30% carbon. The highest grade regularly mined as fuel is anthracite with 90% to 95% carbon; and the final end product of the process is graphite, which is effectively pure carbon and is too valuable to use as a fuel. Traces of graphite are found in metamorphic rocks and this is a clue that buried organic carbon is widely distributed and is far more abundant than its concentrated forms in fossil fuels. The rejection of volatiles in the development of the higher grades of coal is relevant to the recycling of carbon in the environment. The volatiles that have not escaped constitute the now much discussed coal seam gas.

Oil

The starting point for oil is a concentrated community of small organisms, collectively called plankton, typically in a shallow marine environment. As the creatures die and fall to the sea floor, they start to decompose, but are buried in accumulating sediment, so that the decomposition becomes anoxic. If the layer of sediment containing the decomposing plankton is buried to a few kilometres depth and heated to 120°C to 170°C, it may become an oil source rock, 'cooking' the organic matter to produce the raw material for oil. There follows a process, termed cracking in the oil industry, in which large organic molecules are broken into smaller ones, resulting in the 'maturation' of the oil by producing lighter and more fluid material — the lightest component being methane. Since they are more fluid and lighter than the bulk of the source rock, the oil and gas tend to percolate upwards and to concentrate in any porous rock, such as sandstone, and if that is overlain by an impermeable layer with suitable geometry, they collect as an oil and gas reservoir. It is evident from this sequence of events that the development of an oil field requires a very specific combination of geological and tectonic conditions, and that only a small fraction of the starting material will end up as exploitable, trapped oil. Most will escape to the surface in some form, oil seeps or exhaled methane in the case of material developing that far. Expressed as an equivalent global layer of coal, the carbon in the world's known successfully trapped oil and related natural gas deposits would amount to a layer 2.5 mm thick.

Variations of coal and oil development with time and latitude

Figures 12.1 and 12.2 present indirect evidence of the environmental conditions involved in the formation and natural storage of coal and oil. The formation of the first oil deposits 200 million years before the earliest coal is consistent with the earlier development of life in the oceans than on the continents. But most of the exploitable coal was formed about 300 million years ago, quite soon after the earliest deposits, and, although it is a continuing process, the later development has never matched that. The most plausible explanation is that particular fungi, which cause decomposition of lignin, a component of wood that helps to preserve it, had not appeared 300 million years ago. By contrast, although the production of trapped oil began before the first coal, it has been very irregular; the oldest oil is rare and later oil more abundant. A reason for the younger age distribution of oil is that oil (and gas) leak to the surface, as evidenced by the fact that many apparent oil traps are found to be dry. This introduces a cautionary note to ideas about CO_2 sequestration. Figure 12.2 shows the latitudes at which the fuels were formed, which are, in many cases quite different from the latitudes where

(a)

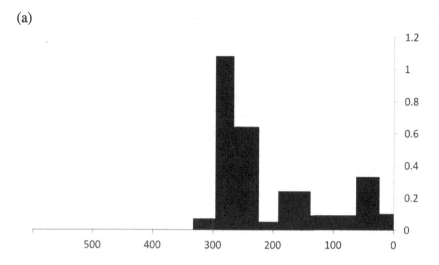

(b)

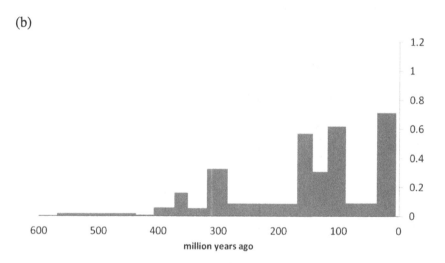

million years ago

Fig. 12.1. (a) The development of global coal reserves over time, represented by shaded areas proportional to the totals for different time intervals. Note that the time intervals are unequal and that the biggest reserves accumulated relatively quickly. (b) Oil and oil-related natural gas reserves from source rocks of different ages. As in (a), the shaded areas represent total amounts for unequal time intervals, which do not coincide with those in (a). Data from Klemme and Ulmishek [12.3].

they are now found. Coal originated mainly at high latitudes, but oil in the tropics and sub-tropics. The virtual absence of coal from the climatic dry belts is readily understood, but the preponderance of very high paleolatitude deposits suggests either a major climatic difference or relative ineffectiveness in cool regions of decomposition processes that destroy dead vegetation. The lack of high-latitude oil suggests that only low-latitude life forms are suitable for generating it, although high-latitude oceans have abundant plankton and might have been expected to produce just as much.

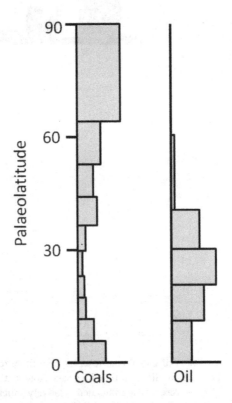

Fig. 12.2. The latitudes at which coal and oil deposits formed. As discussed in Chapter 6, crustal masses have moved and the latitudes at which they occur now do not, in general, coincide with the latitudes of formation (paleolatitudes). After Tarling, [12.4].

Natural gas and coal seam gas

Methane is continuously leaking to the atmosphere from a wide range of sources, which can be traced back to biological origins of carbon [12.5]. It is a major part of the global carbon cycle by which photosynthetically fixed carbon is returned to the atmosphere. The surface decomposition of decaying vegetation produces CO_2 if oxygen is available, but CH_4 in anoxic environments, such as stagnant bogs (where it is known as marsh gas). We have mentioned that methane escapes from coal seams, but more significant is its escape from oil and gas fields, from which the diffusion of natural gas, which is largely methane, produces localised atmospheric concentrations. Much (probably most) of the oil and gas that is generated is not successfully trapped. Gas and oil seepage is rapid enough to indicate that many reservoirs would be exhausted in times that are short compared with their ages and that they are continuously replenished (but not on the time scale of human activity). There may be no traps that are completely tight and geological situations that provide even good ones are rare, reinforcing our cautionary note on plans for CO_2 sequestration. But the escape of fugitive methane is simply completing a carbon cycle, being oxidised to CO_2 in about 10 years, and does not influence estimates of the net burial of reduced carbon that we need for a conclusion about the origin of atmospheric oxygen.

Most of the natural gas that is extracted and used is associated with oil fields. For many years it was discarded as a waste product in the extraction of oil, but its value as fuel is now well recognised, making it an exploration target for its own sake. Less conventional sources of gas may be widely distributed in shales and sandstones of low permeability, from which it can be released by fracturing (fracking), but is not, at least yet, a major contributor. In Table 14.2 gas is identified as providing 23% of the total global energy use and its share is increasing. That share will probably increase faster as plans to tap into coal seam gas develop. This has always been recognised as a serious hazard in underground coal mining and is exhausted to the atmosphere by ventilation For many mines this fugitive gas is a lost fuel source, but the amount that remains trapped in coal seams is much less than the total that was produced during development of the coal and exhaled long ago. The interest in exploiting coal seam gas arises from the fact that it can be extracted from seams which are, for various reasons, not mineable and have not been included in the inventory of fuel reserves, but are far more extensive than those that could be extracted economically as coal. Potentially it expands the estimate of accessible natural gas by a factor of at least 6. This is the incentive to overcome the environmental problems that the coal seam gas industry faces. However, the principal focus of the present discussion is the total amount of carbon in the crust, and the carbon in the methane of even very gassy coal is less than 1% of the carbon in the solid. It does not materially influence our estimation of the total crustal carbon.

Methane clathrate

Mention of methane leakage draws attention to another store of carbon: methane clathrate. This is an interesting compound of methane and water, which chemists have known about for more than 100 years, but its natural occurrence in the crust is a more recent discovery. It is an ice-like solid with a few alternative crystal forms in which the cavities in a framework of water molecules are occupied by other molecules, in this case methane. Clathrate forms when methane under pressure meets water at low temperature, close to its freezing point. Methane clathrate has been identified in sediments down to depths of several hundred metres in several areas, both on the sea floor and on land under permafrost. At least some of it is the result of methane leakage into cold water under pressure, but some appears to be a product of more local methanogenic bacteria. Methane clathrate is unstable if exposed to temperatures appreciably above 0°C or to reduced pressure, and has been a subject of alarmist hypotheses of a catastrophic release, driven by and amplifying global warming. Estimates of the total mass cover a wide range, but the total carbon content could be a million gigatonnes (10^{18} kg), exceeding all the accessible fossil fuels and inviting thoughts about mining it as a fuel source, although the deposits are so scattered that it is not easy to see exploitation as an economic proposition. But, in the present context, methane clathrate appears to be sufficiently abundant to take into account in our estimate of the global total of buried organic carbon, for the purpose of assessing the corresponding oxygen production, and we include the 10^{18} kg clathrate estimate in the total.

Early photosynthesis: blue-green algae

Cyanobacteria, also known as blue-green algae, occur as various widely distributed shallow water species, with a history extending back to the earliest evidence of life on Earth. They are photosynthetic and have been credited with the production of the first free oxygen, but that must be attributed to the hydrogen loss mechanism discussed in Chapter 7. Cyanobacteria occur in both fresh and salt water and are seen as blue-green algal blooms in the water surface, according to seasonal and meteorological conditions. Some types produce toxins and can be a hazard to fresh water supplies. The clearest evidence of their very early existence is seen in fossil stromatolites, aggregates of cyanobacteria which trap drifting sedimentary grains and form solid masses. Although they were more abundant billions of years ago, modern forms are found in many places (Fig. 12.3). In the present context their significance is their photosynthetic activity, fixing carbon and releasing oxygen, but, in the development of early life, they also had a role as nitrogen fixers [12.6]. Fossil evidence of cyanobacteria, as stromatolites or otherwise, is not sufficiently abundant to suggest that their contribution to the free oxygen (or to the formation of banded ironstones) is more than marginal. In any case, in a calculation that follows it is included in the estimated total abundance of carbon in the crust.

The total buried organic carbon

Even acknowledging that the world inventory of fossil fuels may have been underestimated, there is no way that they represent more than a tiny fraction of the buried carbon needed to explain the oxygen released to the atmosphere. A much larger mass of reduced carbon is distributed through

Fig. 12.3. Stromatolites in Hamelin Pool, Shark Bay, Western Australia. Photograph provided by Tourism Western Australia.

the crust. Estimates of its abundance rely on determination of the carbon contents of typical rocks, with care to ensure that they are representative and that their total volumes are reasonably well known. The most important are sedimentary rocks that have not been subjected to substantial heat or pressure. Some carbon survives metamorphism and is found as flakes of graphite in metamorphic rocks, but much of it escapes as carbon dioxide or methane when sediments are heated, and even less reaches the stage of being incorporated in igneous rocks. The total of reduced carbon in the crust is typically estimated as 15 million gigatonnes (1.5×10^{19} kg) [12.7], but this estimate does not include the clathrate, which we add to give a total of 1.6×10^{19} kg. This corresponds to a global coal layer 32 m thick, and is 35 times the carbon needed to explain the present atmospheric oxygen. As evidenced by the appearance of diamonds, the mantle also contains carbon, some of which was probably carried down into the depth range of diamond stability by subducting crust, but we assume that most of the mantle carbon has always been there and does not need to be included in the carbon budget of the surface processes that we are considering.

The carbon dioxide/nitrogen ratio: a comparison with Venus and Mars

We return presently to the estimate of photosynthetically produced oxygen, but first consider the implications for atmospheric CO_2. The carbon in the carbonate rocks of the crust exceeds the buried organic carbon and accounts for most of the now sequestered CO_2, but does not enter the oxygen calculation. Carbonate rocks amount to about 2.1% of the crust and their carbon content is 7.0×10^{19} kg (70 million gigatonnes). We add this to the sequestered organic carbon estimated above, 1.6×10^{19} kg, making a total of 8.6×10^{19} kg, which corresponds to 3.2×10^{20} kg of CO_2. This is 60 times the mass of the present atmosphere. It is a little less than the CO_2 in the Venus atmosphere, but much more than on Mars, and is a plausible estimate of the total CO_2 that the Earth has had, if not all at the same time. We consider also the implication for atmospheric nitrogen, N_2, because, although the total abundances of the two gases are very different, the ratios, CO_2/N_2, are similar for Venus and Mars (by mass: Venus 43.3, Mars 55.4, by Table 7.1). The atmospheres of terrestrial planets must have arisen at a late stage of planetary accretion, when the planets had become massive enough to hold atmospheres gravitationally. Therefore, the gases could not have accreted as gases, but were primarily outgassing products of the solid planetary material and would have been similar for all of the terrestrial (inner) planets. So, we can presume that, if the Earth had no sequestered carbon (or nitrogen), its CO_2/N_2 ratio would be in the same range as is now seen on Venus and Mars. To make the comparison, we consider not just the nitrogen in the atmosphere, but the nitrogen that has been interred with organic carbon. Plants incorporate nitrogen in their chemistry and this is sequestered with the organic carbon. We add an estimate of its abundance from measurements of nitrogen in sedimentary rocks [12.8], which gave a global total of 1.0×10^{18} kg, to the atmospheric nitrogen, 4.0×10^{18} kg, including nitrogen dissolved in the oceans, to give a total of 5.0×10^{18} kg. With our estimate of CO_2, it gives a CO_2/N_2 ratio of 64. Being higher than the values for Venus and Mars, it suggests that, if our numbers are in error, it is because they overestimate the carbon that is buried in various forms. If so, the estimate of photosynthetically produced oxygen that follows from it is likely to be too high rather than too low.

The fraction of atmospheric oxygen derived from photosynthesis

If the oxygen released by photosynthesis were entirely derived by extracting the C from CO_2, then the estimated mass of buried organic carbon, 1.6×10^{19} kg, would correspond to 4.3×10^{19} kg of oxygen, 8.1 times the mass of the atmosphere and 35 times the present atmospheric oxygen. However, not all of the oxygen comes from CO_2. As in the example on page 189, the end products of the burial of organic carbon are hydrocarbons, meaning compounds of carbon from CO_2 and hydrogen from water, with oxygen being released from both. For an estimate of the total photosynthetic oxygen, the amount released from CO_2 must be multiplied by a factor between 1 and 2, 1 for the highest grade of coal, with negligible hydrogen, and 2 for methane, the lightest hydrocarbon. The appropriate average factor depends on the relative abundances of the various hydrocarbons. There are numerous studies reporting the hydrogen to carbon ratios of fossil fuels by numbers of atoms, almost all at the low end of the possible range of 0 to 4 (graphite to methane). For solids the ratio is typically 0.6, only a little higher for oil and highest for gas, but still well below the pure methane value. The highly dispersed carbon, which dominates the crustal content of organic carbon, is in the form of degassed solids, with the ratio well below 0.6. Allowing for a possible greater abundance of gas than has been fully recognised, especially with the clathrates, we assume a factor of 0.7 for the crustal organic H/C ratio, which is probably generous. On this basis the photosynthetic oxygen exceeds the quantity derived from CO_2 by the factor 1.175, raising the estimated production in the life of the Earth from 35 to 41 times the present atmospheric level.

With the estimate in Chapter 7 of the oxygen consumed by weathering, photosynthesis can explain about 40% of the total oxygen production. This agrees with the estimate of 60% released by the photolysis of water vapour and escape of hydrogen. But the two contributions to atmospheric oxygen have varied quite differently with time. The hydrogen loss mechanism would have been almost steady for the whole life of the Earth. Photosynthesis began later and started slowly, but is now making a bigger contribution. We consider hydrogen loss to have provided a 'base load' supply, but that it has been almost entirely consumed by weathering, most obviously when the continents were bare, and that abundant atmospheric oxygen appeared only when life on land became a major source. However, it appears that, although photosynthesis dominates the present oxygen production, it would not have developed without priming by the hydrogen loss mechanism.

HUMAN INFLUENCES

13. Effects of fossil fuel use

The current rapid rise in atmospheric CO₂

A 50-year record of the atmospheric CO_2 concentration obtained at the Mauna Loa observatory on Hawaii Island is plotted in Fig. 13.1(a). It shows a consistent annual oscillation of peak-to-peak amplitude about 6 ppm, driven by the seasonal growth and decay of plant life in the northern hemisphere, superimposed on the longer term increase in CO_2 caused by fossil fuel burning. Ignoring, for the moment, the long-term trend, the process is cyclic. Carbon fixed during the summer months is either eaten by animals (including humans), that breathe out the CO_2, or gradually decomposes, completing the carbon cycle. A similar, slightly shorter record from the southern hemisphere has been obtained at Cape Grim, on the north-west tip of Tasmania, Australia. It shows a much weaker annual cycle, as expected for the smaller land area in the southern hemisphere, with a peak in October, instead of April, and values that are slightly lower than the data from Hawaii. The difference in the annual averages, evident in the figure, is a consequence of the fact that most of the fossil fuel use is in the northern hemisphere, and the transfer of atmospheric constituents across the equator is slower than redistribution within hemispheres. This was recognised during atmospheric testing of nuclear weapons. Identification of fossil fuel burning as the cause of the strong upward trend in atmospheric CO_2 was obtained many years ago by measurements of the ratio of carbon isotopes. There are two isotopes of interest. The common one has an atomic mass of 12, with a rarer one of mass 13 that is slightly less

(a)

(b)

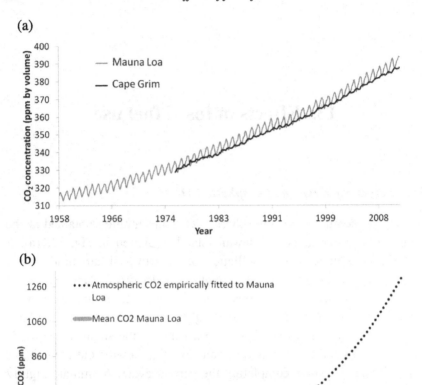

Fig. 13.1. (a) Atmospheric CO_2 concentration recorded at the Mauna Loa observatory on Hawaii Island and at Cape Grim on the north-west tip of Tasmania, Australia. The values are presented as parts per million by volume. For mass fractions multiply by 1.5. Mauna Loa data are from NOAA Earth System Research Laboratory (ESRL) and Cape Grim data are from CSIRO and Australian Bureau of Meteorology Cape Grim Air Pollution Station. (b) A simple mathematical extrapolation [13.1] of 12-month running means of the Mauna Loa data in (a). This is a 'business as usual' extrapolation, which assumes no serious control of fossil fuel burning and no change in the manner of absorption by the oceans.

abundant in vegetation than in the atmosphere in which it grew. The fossil fuels, being biological products, have the biological isotopic ratio, with low ^{13}C, and the proportion of this isotope in the atmosphere is progressively decreasing as fossil carbon is added. A possible bias by preferential solution of ^{13}C in the ocean would still be too small to influence this observation.

As we demonstrate later in this chapter, the increasingly rapid upward trend in Fig. 13.1(a) is a new phenomenon, with no parallel for at least the last million years of the geological record. This is a matter of particular environmental concern and has led to numerous extrapolations, one of which is shown in Fig. 13.1(b). There is no unambiguous or agreed theory for this curve; it is simply an empirical fit to the Mauna Loa data for 1960–2010 [13.1], with extrapolation based on the 'Business as Usual' assumption that the trend of the last 50 years will continue. The use of fossil fuels is not changing fast enough to cause much doubt about the extrapolation to 412 ppm in 2020. Further extrapolations to 450 ppm (an 'allowable' limit suggested in IPCC reports) in 2034, to 504 ppm in 2050, a common reference date, or to 560 ppm, twice the pre-industrial value, in 2064, are more questionable because the resolve to curb CO_2 emissions may by then have had some effect, but also because of uncertainty in the role of the oceans in dissolving CO_2, and because global warming may release CO_2 and methane from permafrost. As fossil fuel use increases, the consequences become more obvious, and therefore more widely recognised, but they have been understood scientifically for many years. Budyko [13.2] documented numerous reports predating the 1979 (Russian) edition of his book, drawing attention to climatic effects to be expected from increased atmospheric CO_2.

Solution in the oceans

The accelerating rate of increase in atmospheric CO_2, seen in Figs. 13.1, had reached 2.14 ppm/year in 2010, but this is much less than the rate of release of CO_2 by human activity. Adding deforestation and cement production to fossil fuel use, this rate was about 35 billion tonnes per year in 2010 [7.2], corresponding to 4.36 ppm/year by volume for the atmosphere, almost precisely a factor of two more than the observed atmospheric increase. The difference must be attributed primarily to solution of CO_2 in sea water, with a minor contribution by its role in the weathering of silicate minerals. There is a continuous exchange across the sea surface with a net transfer that depends on the departure from equilibrium between the concentrations in the air and water. Although equilibrium must have prevailed globally during the steady pre-industrial era, and we can attribute the present net rate of solution to the departure from that state, it is not a simple balance between the concentrations. CO_2 is more soluble in cold water than warm water, with the result that there is a net release by the equatorial ocean and net solution at high latitudes. As discussed in Chapter 6, cold, saline polar water sinks to the ocean floor, and it takes some of the CO_2 with it. Equatorial water flows pole-wards (accounting for a substantial transfer of heat, as noted in Table 8.2), the cycle being completed when the deep water reaches the surface at low latitudes and releases CO_2 as it warms up. The process is slow and cannot maintain even an approximate balance with the present rate of increase in atmospheric CO_2. So, the excess CO_2 is now dissolving in the shallow water, possibly at all latitudes, even the warm tropical seas.

The geographical variability of the uptake of CO_2 by the oceans is such that it would be very difficult to make a direct observation of the total effect. From the limited available data the assumption that the fraction of 'new' CO_2 that has dissolved has been constant appears reasonable, that is, half of the emitted CO_2 is dissolving in the oceans. At first sight this may not appear important because the oceans hold something like 50 times the atmospheric content and the fractional change is correspondingly smaller, but that is misleading. In any case, exchange with the deep oceans is very slow and only the capacity of the shallow water is relevant on the time scale of our interest. We need to consider just what solution in the ocean really means. It is not simply a case of mixing CO_2 gas and sea water; the chemistry is a very sensitive balance between CO_2 gas, carbonic acid, calcium ions, calcium bicarbonate and calcium carbonate [13.3], with temperature, pressure and ocean acidity having controlling influences. In normal (pre-industrial) times, the solution of calcium from eroding rock and hydrothermal activity sufficed to maintain a balance but, as concluded in Chapter 6, it can no longer do so and the result is increasing acidity. Not only does this reduce the ability of the ocean to take up more CO_2 but, if unchecked, it will inhibit the continued growth of carbonate shells, coral etc., and eventually even dissolve limestone, with a net release of CO_2 into the ocean-atmosphere system. The consequent changes to the marine environment appear as threatening as does global warming to the land environment.

Ice core records of ice age CO_2

A longer history of the atmospheric abundance of CO_2, before there were direct measurements, is obtained from air trapped as bubbles in Antarctic and Greenland ice and extracted from deep cores. From areas with very rapid deposition of snow, for which the diffusion of air within unconsolidated snow (firn) can be allowed for, it is possible to obtain values of atmospheric CO_2 concentration for the historical period. The measurements show that, for more than 1000 years before the industrial era, the concentration was reasonably steady at about 280 ppm (by volume), with some minor fluctuations that appear to be climatically related. For modern times, we adopt 280 ppm as the baseline and attribute the increase above that to fossil fuel burning. Much longer records require long cores of slowly deposited ice, for which the diffusion problem precludes the possibility of inferring fine details. However, they show the longer-term changes that occurred through a series of ice ages, extending back almost a million years. Figure 13.2 summarises data from the Vostok core, obtained at the Russian base in Antarctica. The heavier line shows a quasi-cyclic oscillation in the CO_2 content of air bubbles trapped in the ice over the last 400,000 years (parts per million by volume, as for Fig. 13.1). For all this time the concentration was within the range 190 ppm to 290 ppm. The broken line on the same graph is a record of variations in temperature estimated from

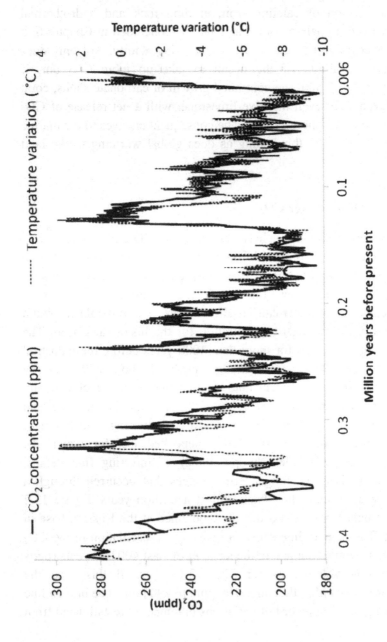

Fig. 13.2. 400,000-year records of atmospheric carbon dioxide and temperature from the Vostock, Antarctic, ice core, plotted from data collated by CDIAC (Laboratoire de Glaciologie et Geophysique de l'Environment, France, and Arctic and Antarctic Research Institute, St Petersberg, Russia).

the deuterium/hydrogen (D/H) ratio in the ice. The annual cycle of snow precipitation, apparent in the layering, fixes the time scale.

There are several features of this striking record that draw attention to the relationship between CO$_2$ and temperature through four glaciation-deglaciation cycles. The CO$_2$ – temperature correlation is very obvious and invites attempts to look at the fine details to see what inferences can be drawn about cause-effect relationships. But there are several time lags in both the temperature and CO$_2$ variations that frustrate any attempt to infer directly from the graphs that one drives the other. An interpretation that we hint at in Chapter 10, and pursue quantitatively presently, is that the temperature variation was driven by the 96,000-year cycle of orbital eccentricity (one of the Milankovitch cycles discussed in Chapter 3), and that the CO$_2$ changes were a consequence. But first we point out that the interpretation is not straightforward for either CO$_2$ or temperature variations. Several considerations need to be taken into account:

(i) As mentioned above, when snow first settles it is a loose aggregation of ice crystals through which atmospheric gases readily diffuse. In time, with an increasing over-burden of snow, it is compressed to ice, but deep burial (50 to 100 m) is required to consolidate the ice sufficiently to enclose the gases within sealed pores and, in slowly accumulating ice, that takes several thousand years. Thus, the composition of the gas that is trapped at any particular level is a compromise between the compositions at the time of snow settling and both earlier and later times from which gases have diffused within the snow — biased to later times because the gases diffuse more readily in the later, less compressed snow. This means that the gas is younger than the ice in which it is encapsulated, accounting for at least part of the time lag apparent in Fig. 13.2.

(ii) The temperature variations in Fig. 13.2 represent repeated glaciations (ice ages) and deglaciations. Although it is atmospheric temperature that is recorded, the changes are slow enough for ocean temperatures to stay closely in step, so the record represents changes in global ocean temperature. As mentioned, there is more CO$_2$ dissolved in the oceans than held in the atmosphere, but its solubility in sea water depends on temperature. Thus, the partitioning of CO$_2$ between ocean and atmosphere varies with the global temperature. During cold periods more CO_2 is dissolved, reducing the atmospheric content, and it is returned to the atmosphere when the oceans are warmed. But the oceans have a high heat capacity, heating up or cooling down slowly in response to atmospheric temperature changes. Thus the solution or rejection of CO_2 is delayed relative to the atmospheric temperature changes.

(iii) Temperatures were obtained from the D/H (deuterium/hydrogen) ratio in the ice. In the partitioning of these isotopes between water vapour and ice or liquid water, deuterium favours the solid or liquid and is relatively less concentrated in the vapour. The difference diminishes with increasing temperature, so that the deuterium concentration in ice is a measure of the temperature at which it condensed if the concentration in the vapour is known (or assumed to be constant). The same D/H partitioning operates in the evaporation of water from the ocean. Atmospheric water vapour is isotopically lighter than ocean water. When the vapour condenses, as rain or snow, this partitioning is reversed by the preference of deuterium for the solid or liquid, but only partially so. The fraction of the vapour that condenses is much larger than the fraction of sea water that evaporates. In the limit of complete condensation there can be no isotopic separation at all and polar conditions come close to this. For this reason isotopic partitioning is weaker in condensation than in evaporation, so that it is not just atmospheric water vapour but also rainwater and polar ice that are isotopically lighter than sea water. But it does not matter whether the ice D/H ratio records the temperature of evaporation or precipitation, as long as it always records the same thing, because our interest is in comparing temperatures at different times. However, there is a further question to consider.

(iv) During ice ages (the minima in Fig. 13.2) large volumes of ice are locked up in polar regions, amounting to as much as 3% of the ocean water. Since the ice is isotopically light, the remaining ocean water is enriched in deuterium and there is a corresponding increase in the deuterium in water evaporated from it. This then appears in the ice precipitated in polar regions and the deuterium enrichment corresponds to a lower temperature than actually prevails, exaggerating the estimated change in temperature, relative to the warmer interglacial periods. But the error is small, both because the deuterium enrichment arises from the temperature dependence of D/H partitioning between water and vapour, which is not very different from the partitioning between ice and vapour that occurs in the precipitation, and because the oldest ice was precipitated when the ice caps were smaller. The error in calculating the temperature change corresponds approximately to the fraction of ocean water sequestered as ice, no more than about 3% or 0.3° in the 10° temperature change in Fig. 13.2. This is less than other uncertainties in the observations.

There remains a doubt about just what the measured temperatures refer to. They must be some compromise between the temperatures of evaporation and precipitation, with complications arising from extended evaporation and partial precipitation on the way to final precipitation as polar ice. The temperature changes were greatest at high latitudes, so the measured ice core temperatures overestimated the changes at lower latitudes, being biased by the dominance of high latitude ocean in controlling the isotopes in the ice. Although the ice core temperatures exaggerated the global temperature changes, we take the 10° temperature changes in Fig. 13.2 as a guide to a quantitative understanding of the CO_2 – temperature relationship through the ice ages. This leads us to a crucial conclusion.

Ice age CO₂ interpretation

The CO_2 concentration in Fig. 13.2 has oscillated between two approximate limits that differ by a factor 1.4, 280 ppm and 200 ppm, with the low concentrations occurring at times of low temperature. The sensibly constant values of the limits, and the fact that the upper one coincides with the well-determined pre-industrial average concentration, invite the inference that there was no large input of 'new' CO_2 during this period and that it was simply exchanging between the atmosphere and ocean. This is what would be expected qualitatively if more of it dissolved in the oceans when they were cold. A simple calculation gives

Stewart Callendar, 1898–1964
British steam engineer with a strong interest in meteorology, who concluded that global temperature was rising and, as early as 1938, asserted that this could be attributed to increasing atmospheric carbon dioxide. In spite of doubt that his data were adequate to support the claim, he maintained his position and must be credited with drawing to attention a problem that is now accepted as very significant.

Charles Keeling, 1928–2005
American chemist, responsible for developing the first reliable and accurate method of measuring the concentration of carbon dioxide in the atmosphere. The measurements which he initiated at the observatory on Mauna Loa, Hawaii, have been maintained continuously for more than 50 years, yielding data represented by a graph in Fig. 13.1, known as the Keeling curve. This work is now overseen by his son, Ralph Keeling.

quantitative confirmation. In the temperature range of interest, the solubility of CO_2 in sea water increases by 4.4% per degree fall in temperature. This means that, because the ocean capacity is much greater than that of the atmosphere, if the balance in the partitioning of CO_2 between them is maintained as the temperature changes, then the atmospheric concentration falls by 4.4% for each 1°C fall in temperature. A concentration change by a factor 1.4 requires a temperature change of 7.8° [13.4], somewhat less than the temperature range plotted in Fig. 13.2, but consistent with an effective average temperature change for the ocean water that evaporated to yield polar ice. As nearly as can be determined, the total amount of CO_2 in the atmosphere and oceans has been constant for the whole length of the record, encompassing four cycles of ice age advances and retreats. Contributions by volcanic eruptions and other disturbances have been accommodated by the natural balance. A greenhouse effect must have occurred, reinforcing the observed changes, but it was not the driver of them. Overriding the discussion of details in Fig. 13.2, two conclusions are emphatically clear: (i) There is no precedent in this extended record for the increase in atmospheric CO_2 in the last century or so, and (ii) Over the whole time of the record, hundreds of thousands of years, the total CO_2 content of the atmosphere and oceans has remained constant. The natural sources and sinks were balanced and the atmospheric variations were a consequence of the temperature dependence of CO_2 solubility in the oceans. Although, for the duration of the ice age record, the atmospheric CO_2 concentration was driven by temperature variations and was not the cause of them, it would have provided feedback by greenhouse reinforcement. The significance of this conclusion, pursued in Chapter 15, is that the industrial era CO_2 emissions are effectively a permanent addition to the atmosphere-ocean system and greenhouse warming will be superimposed on whatever natural temperature changes occur.

The oceans delay the effects of greenhouse changes

The evidence of CO_2 partitioning between the atmosphere and oceans when the total remained constant but temperature varied, offers some insight on what to expect from the progressive increase in CO_2 with fossil fuel burning. The oceans dissolve more of it as the atmospheric concentration rises, but the extent to which they do so will decrease as temperature rises, and there will inevitably also be chemical limitations. But the changes seen in the ice age record of Fig. 13.2 were very slow compared with what is happening now. The peaks and troughs of that record mark equilibrium states that take at least hundreds of years and probably thousands of years to establish. Thus, the ice core observations record the equilibrium partitioning of CO_2 between the atmosphere and oceans at different temperatures, but the present situation is a transient one, with changes occurring too fast for equilibrium to be established. Although the oceanic fraction of the global CO_2 is reduced by rising temperature, it is large enough to ensure that, under equilibrium conditions, for any modest temperature rise, much of the fossil fuel CO_2 that is released into the atmosphere by human activity dissolves in the ocean. The time that it takes to do so introduces a delay in the approach to equilibrium, but, as far as surface temperatures are concerned, the thermal lag arising from slow heating of the oceans is more significant.

The transience of the ocean heating is discussed in Chapter 8, where we point out that, although an important part of the current rate of sea level rise is attributed to thermal expansion, in the longer term it can be no more than a minor contribution to the rise that must be expected. Table 8.2 gives estimates of the current net heat inputs to several components of the Earth, with the ocean component dominant. This is necessary for significant thermal expansion, because the expansion coefficient is low but the heat capacity is high. Allowing for uncertainties, the heating of the oceans accounts for 85 to 90% of the net heat input to surface features of the Earth. Even ice cap melting is included in the other 10% or 15%. When equilibrium is approached and the oceans stop warming, all the incoming heat will be radiated away, requiring a higher surface temperature. But before such a balanced or equilibrium state is reached, for a reason discussed in Chapter 15 much of the greenhouse induced heat will be diverted from the oceans to the ice caps, accelerating melting and sea level rise. An equilibrium state means not just a balance of the temperatures and CO_2 contents of the atmosphere and oceans, but also a stable state of the ice caps (or lack of them) and that may be slow to establish.

Only complete cessation of emission can solve the CO_2 problem

At several points in the discussion we have referred to an equilibrium state, from which the Earth is now well removed, and we consider what it means in the light of the conclusions in this chapter. It requires a constant mean surface temperature that causes all of the greenhouse-induced heat to be radiated away, with no added heat stored in the oceans or elsewhere, and for this the atmospheric CO_2 concentration must remain constant. But we see from the ice age data that an equilibrium state means that the total CO_2 of the atmosphere and oceans together will remain constant, so there can be no continuing solution in the oceans and therefore no 'new' CO_2. An equilibrium state means that there can be no continuing increase in CO_2 at all. The notion of a stable state with any fossil fuel burning (unless it is fully compensated by sequestration) is illusory. There can be no 'allowable' emission that would avoid an indefinite build up, with a continuing temperature rise and its consequences. All that 'controlled emission' could achieve is a delay in approaching the ultimate extreme greenhouse effect.

Proponents of CO_2 sequestration argue that our conclusion is not unavoidable because it would be possible to capture all emitted CO_2, compress and liquefy it and inject it into exhausted gas wells and other sealed geological reservoirs to be trapped eventually by structural, dissolution, residual or mineral mechanisms. While it is possible to sequester some CO_2, that would still leave us with 'controlled emission', which, as we have pointed out only delays the inevitable. Sequestration can be justified as an interim measure, during lead up to full deployment of alternative energy sources, but that is all. If sequestration of all CO_2 were to be effective in avoiding an indefinite build up, it would require compensating capture of CO_2 from the atmosphere, but that faces the combined difficulties of a high energy demand itself and the fact that the requirements for suitable reservoirs are sufficiently demanding to ensure that there is, in principle, capacity for only a small fraction of the CO_2 currently being produced. So, we are sceptical and regard sequestration as offering little more than a short-term political solution to the basic problem, not a scientific one, providing an excuse for delaying serious action in promoting 'renewable' energy sources [13.5].

A thought about health effects

Our attention has concentrated on the greenhouse effect of increasing CO_2 and the consequences of that, but we must wonder what other implications there are. Our identification of hydrolysis of water and hydrogen loss as a major source of oxygen (Chapter 7) breaks any assumed nexus between the development of an oxygen atmosphere and the removal of CO_2. Oxygen is a sufficiently large component of the atmosphere to ensure that its abundance is not endangered by fossil fuel burning. The concern is that most familiar life forms are conditioned by the restriction of CO_2 to a trace abundance and we are witnessing a large percentage change. Even if atmospheric oxygen would have become adequate for familiar advanced life forms 1.8 billion years ago, this did not coincide with the reduction in CO_2 to the level of a trace constituent. Removal of what is, to us, excess CO_2 probably continued until 500 or 600 million years ago, and perhaps even later, by which time it was approaching a concentration suitable for creatures such as ourselves. It appears possible that the rapid development of fossilisable animals after about 500 million years ago had as much to do with the decrease in atmospheric CO_2 as with the increase in oxygen. It is not clear what is the highest permanent level of CO_2 that would be biologically tolerable. All observations are of the effects of short-term exposures, for which a limit of 2% (20,000 ppm) has not been reported as harmful. The health effects of even slightly higher concentrations increase strongly with the duration of exposure and it appears unlikely that anything like 2% would be acceptable in the long term. But this is well above the level that can be expected from even the most extreme fossil fuel use. The only danger appears to be accidental exposure to high concentrations, arising from sequestration processes that could be quickly lethal. The possibility was disastrously demonstrated in 1986 when 1800 people were asphyxiated by a sudden discharge of CO_2 from Nyos, a volcanic lake in Cameroon, West Africa [13.6]. This event drew to attention the danger of high CO_2 concentrations that can arise naturally, and is a warning that considerable care will be needed to ensure that sequestered CO_2 is safe, and especially that it cannot leak into lakes or tapped groundwater.

14. A comparison of human energy use with natural dissipations

Almost all available energy is ultimately derived from the Sun

In our preface we anticipate the discussion in this chapter, because it is the use of energy that rings the loudest environmental alarm bells. But when fossil fuels are exhausted, or embargoed because of their environmental impacts, there will still be plenty of energy available. Solar energy has maintained life and much else for hundreds of millions of years and can do so for at least as much longer. It has always been the primary energy source for the Earth and will remain so. In this chapter we compare the energies of various natural effects, with the obvious conclusion that solar energy is much greater than anything else (Table 14.1), and that wind is the only other accessible energy exceeding the human use of energy. This is emphasised by considering not just current energetic processes, but those occurring over the entire life of the Earth, as summarised in Table 5.1. The solar energy received and re-radiated back into space would, in 4.3 million years, be equal to the total stored energy in the Earth or, in 70 million years, to the gravitational energy of its formation [14.1]. Almost all of our exploitable energy sources are derived from solar energy and are, in that sense, secondary sources. Tidal, geothermal and nuclear energies are exceptions, but river flow and wind are included and, of course, fossil fuels are fossilised solar energy. Faced with the need to consider 'new' energy sources, it is logical to turn, as closely as possible, to the primary energy and, as we point out, alternatives are very limited.

219

Table 14.1. Some natural energy dissipations. All values in terawatts.

Heat flux from the	global	44.2
Earth	land	9.6
Solar radiation	top of atmosphere	174 000
	surface	98 000
	land surface	28 000
Wind friction	global	>430
	land	>120
Tides	total	3.7
	marine	3.2
River flow		6.5
Breaking waves		~5

The magnitude of the energy problem

The numbers in Table 14.2 indicate the scale of the energy problem. The total rate of all forms of energy use now exceeds 16 terawatts, but we need to look closely at the individual entries in the table to see what this means if fossil fuel use is to be seriously reduced. The listed energies for the fossil fuels refer to the thermal energy released by complete combustion. A large fraction is used for electricity generation, especially in the case of coal, and we need to assess that fraction differently from the other entries in the table. Under ideal conditions, modern power stations operate at efficiencies up to about 40%, limited by thermodynamic principles [14.2] as well as technical questions. However, the world average efficiency is only about 31%, that is 31% of the thermal energy in the fuel is converted to electrical power. The other 69% of the energy in the fuel is discarded as waste heat by power stations. The power classified as secondary in Table 14.2 is not included in the total energy use, because it uses fuel listed with the primary sources and is not additional to them. If this secondary energy is to be replaced by expanding the primary sources (the 5ᵗʰ and 7ᵗʰ entries in Table 14.2 or Table 14.1) the 31% efficiency factor would not apply,

Table 14.2. Global energy use in 2011. All values in terawatts.

Primary sources	coal	5.05
	oil, liquefied gas	5.80
	natural gas	3.83
	wood, animal waste	1.36
	hydroelectricity	0.42
	nuclear power	0.32
	solar, wind, wave, tidal, geothermal	0.05
	total	16.83
Secondary	coal, oil, gas-fired electricity generation	1.49

and the listed (thermal) energies of the primary sources would be reduced by (1.49/0.31 =) 4.81 terawatts, and the total energy by (4.81 – 1.49 =) 3.32 terawatts. This is the 69% of 'waste' heat discarded by power stations. Since most sources of 'renewable' energy [13.5] produce it in the form of electric power, in contemplating the use of them to replace fuel-based power, it is appropriate to rate the present total 'end use' demand as 13.5 terawatts, rather than the 16.8 terawatts in the table. These numbers assume electricity generation in centralised, industrial-scale facilities. This means that before it reaches consumers, it is subject to losses in transmission and distribution, which are typically 7%, and if we discount that, the estimated total demand is reduced to 13.4 terawatts. If the 'traditional' energy sources, wood and animal wastes, are also discounted, then the end use power demand is reduced to 12 terawatts. Nevertheless, that is still a bigger number than several of the natural energy dissipations in Table 14.1 and leads directly to two conclusions: (i) the human use of energy is a global geophysical-scale phenomenon and (ii) the options for comprehensive replacement of fossil energy are limited. We examine, in turn, each of the apparent options indicated by Table 14.1 from the fundamental perspective of their availability in principle. We do not address the technical problems of exploitation.

Geothermal energy

Over most of the continental areas the heat flux from the interior amounts to about 65 milliwatts/m² (65 kW per square kilometre) and the increase in temperature with depth is about 20°C/km. There is little prospect that such low grade heat will be useful. Exploitable geothermal energy is found only in the limited areas of recent or current igneous activity. Although much of the Earth's heat is derived from radioactivity, the radioactive elements are widely distributed and much too dilute to be of interest as a heat source. Granite is the most radioactive of the common rocks and its radiogenic heat is typically 3 microwatts per cubic metre (3 kW per cubic kilometre). If a sample of granite were thermally isolated from its surroundings and allowed to heat up by its own radioactivity, its temperature would rise by 40°C to 45°C per million years. The radioactivity of local rock is irrelevant to the assessment of prospective geothermal sources. The search is for heat from the igneous origins of accessible rocks and that necessarily means that they are geologically youthful.

Established geothermal power stations operate in areas that are now or were recently volcanically active, with very hot rock, or even magma, close enough to the surface for groundwater to be strongly heated. It is drawn off via bores as superheated water, producing steam that drives turbines. Although the technology is well developed in widely separated places, especially Iceland, California, Italy and New Zealand, it requires particular geological conditions — specifically porous sediments, supplied with groundwater, overlying the hot rock or magma. Proposals for more widespread extraction of deep heat depend on ideas for fracturing hot rock and circulating water through the cracks. Granite is a favourite target because it occurs in great masses, termed batholiths, large enough to retain their igneous heat for hundreds of thousands of years. For an idea of the energy derivable (in principle) from a 'hot dry rock' power installation, consider the heat that would be extracted by cooling a cubic kilometre of granite by 100° and using the heat to generate power with 20% efficiency. The energy yield would be 1400 megawatt-years, 100 megawatts for 14 years [14.3], 7 parts per million of the global energy demand for those 14 years. The rock would then be thermally exhausted and would not recover. This would be mining heat, not use of a renewable resource in the conventional sense, but volcanism continues, so we could consider it renewable on a sufficiently long time scale, recognising that it appears in different places, at times separated by millions of years. Unless we include the volcanism at mid-ocean ridges, only a modest fraction of the heat flux listed in Table 14.1 is volcanic, and only a fraction of that appears as hot dry rock. Assuming continuous production, the rate would average about 1 km³/year globally and, if fully utilised, yield a power of about 1.4 gigawatts. This is one ten thousandth of the global energy demand. Since that is not realistic, we must regard geothermal energy as a finite, mineable resource. There is no concern about global geophysical consequences of its use but, independently of technical difficulties, the limited availability excludes it from the big league in the energy game.

Solar energy

Although, as Table 14.1 makes clear, solar energy is abundant globally, use of it is limited not just by technological and economic factors, but by variations in its accessibility with latitude, season, time of day and even weather. But its super-abundance demands that these difficulties be addressed. Whatever use we make of it can have no material influence on it. In any case the human use of 'renewable' energy is not a permanent extraction from the Earth, but a conversion to alternative forms, all of which end up as heat. In the case of solar energy we merely intercept a tiny fraction of it and delay its conversion to heat. Where and when sunlight is abundant, the concept of efficiency, as the ratio of power output to solar energy input, is relevant only so far as it determines the cost of conversion to usable forms, especially electrical power. Sunlight is abundant (solar radiation reaching land is 2000 times the human energy use). This means that the fraction converted is of no concern, although the area required to intercept and use it imposes local constraints.

If solar energy is used to drive steam turbines, it requires strong focussing of sunlight to produce the necessary high temperature. This is inevitably an industrial scale operation. An increasingly popular alternative is the use of photovoltaic cells, the solar panels familiar on many roofs around the world, which are ideally suited to small scale installations, as in residential areas, most obviously in remote rural areas. This reduces the transmission losses (and cost) of electrical distribution from major installations. The efficiencies, controlled by the semi-conducting properties of the cell materials, are progressively improving, but are low by the standards of thermal power generation. However, manufacturing cost is of greater concern than cell efficiency and, if the area requirement imposes no constraint, cells of low efficiency may be the most economic option. The problem of intermittent generation is common to several energy sources (solar, wind and tides), and we refer to it at the end of this chapter.

Wind power

Wind has a long history as a source of mechanical power. Windmills have been in use for at least 3000 years and there is no clear earliest use of sails for boats. But wind power was eclipsed and almost forgotten in fossil fuel-driven industrialisation. With that now called into question, and with continued expansion of the demand for power, the scope for wind as a major contributor needs close scrutiny. The ultimate driving force is the heat of the Sun, which causes atmospheric convection. The Earth's rotation deflects the motion into cyclonic and anticyclonic circulations with turbulence on many scales. The convective driving mechanism requires the whole atmosphere to be in three-dimensional motion; it could not operate in a thin layer because the energy is derived from the buoyancy of density differences over large ranges in elevation. Winds in the upper atmosphere are stronger than near the surface. Partly because the surface is a frictional boundary, wind speed increases with height, and energy of the motion is fed downwards. Most of it is dissipated in turbulent mixing, but some reaches the surface, and the rate at which this occurs is the power fed into a boundary layer near the surface and dissipated there. It is a measure of the energy that can be intercepted by turbines operating at manageable heights in the boundary layer. By estimating the total energy dissipation by interaction of wind with the surface of the Earth, we have an upper bound on the energy that could, in principle, be extracted from it in this way.

Figure 14.1 presents observations that offer a global approach to such a calculation. Small variations in the rate of rotation of the Earth are measured by comparing the signals from remote (galactic or intergalactic) radio sources at widely separated receivers, with the results shown as the solid line in the figure. Changes amounting to a few milliseconds in the length of the day are seen to occur quite rapidly and the only plausible explanation is an exchange of angular momentum with the atmosphere. Validity of this explanation is confirmed by the broken line in the figure, which is derived from global observations of atmospheric motion, with the scale adjusted to give the corresponding effect on the Earth, assuming the total angular momentum of the Earth and atmosphere together to remain constant. The speed of the observed changes demonstrates that the mechanical coupling of the Earth and atmosphere is quite tight. It is this coupling that accounts for the frictional dissipation of wind energy. There is an obvious annual cycle in the figure, but our present interest is in the very rapid variations. How quickly does the Earth respond to changes in atmospheric circulation? In Fig. 14.1 we see changes amounting to a millisecond in the length of the day occurring in about 15 days. We use this as a measure of the coupling to estimate the rate at which frictional contact with the Earth is dissipating the energy of atmospheric motion.

The wind component that relates to the length of day (LOD) variations is the east-west motion of the atmosphere as a whole, which we refer to as the average zonal wind. To cause a 1 millisecond change in the LOD requires the speed of this wind to change by 2.5 metres/second and, for this to occur in 15 days, the corresponding energy dissipation is not less than 13 terawatts [14.5]. But this speed is just the average east-west drift speed of complicated atmospheric motions that are stronger by an average factor that we estimate to be 5.5 or 6.0,

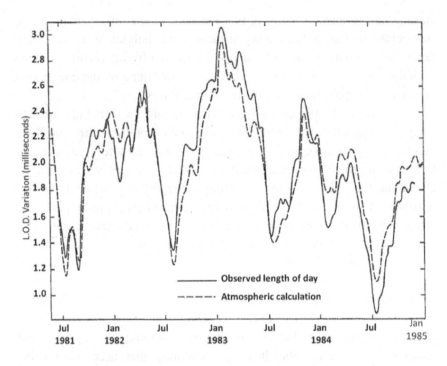

Fig. 14.1. Fluctuations in the rate of rotation of the Earth, determined from precise observations of remote radio sources (Very Long Baseline Interferometry, VLBI), shown as the solid line. The broken line shows the variations that would be expected from meteorological observations, with conservation of the total angular momentum of the Earth plus atmosphere [14.4].

and, since energy varies as the square of wind speed, we arrive at the 430 terawatts of wind dissipation in Table 14.1. This is a very approximate estimate; it could be a serious underestimate if, for example, the apparent response time of the Earth to atmospheric changes is not a measure of the coupling, but is a characteristic of the atmospheric motions. Since it is at least 30 times the human use of energy, wind must be rated as a potential major contributor to renewable energy. The same conclusion was reached from a more immediately practical study of land areas with winds strong enough and consistent enough to be useful [14.6]. For this purpose wind strength at 80 metre elevation was assessed, this being a typical wind turbine height. The conclusion was that accessible wind energy sufficed for the total human energy demand.

It is worth emphasising that, in spite of local objections to wind projects on the basis of noise or the introduction of artificial landscape features, there are no meteorological or global environmental concerns about harnessing wind. It is driven by turbulent convection in an atmospheric layer that is many kilometres thick, with downward transfer of energy and inevitable dissipation in a lower boundary layer. There is no problem in using turbines to extract some of the boundary layer energy that would be dissipated anyway. If we postulate mechanisms so effective that all wind is stopped in (say) the bottom 100 metres of the atmosphere, all that would do is raise the effective lower boundary of the turbulence by 100 metres, with the atmosphere otherwise unaffected. The total energy of atmospheric processes is vast and energy dissipation in the body of the atmosphere dwarfs that in the boundary layer. There can be no concern about harnessing some of the boundary layer energy. We are really considering an aspect of solar energy that is first harnessed naturally by the atmosphere and can then be intercepted as wind. When used it becomes heat, which is what would happen to it without human intervention, so that it satisfies our definition of renewable energy [13.5].

Tides

As discussed in Chapter 2, the dissipation of rotational energy by tides has had a major effect on the rotation of the Earth and the orbit of the Moon. But the present rate of dissipation appears anomalously strong and in Chapter 6 we attribute this to tidal resonance of the Atlantic Ocean. Without this resonance the total tidal dissipation would be nearer to 1.3 terawatts than to the 3.7 terawatts listed in Table 14.1, with about 0.8 terawatts attributable to marine tides instead of 3.2 terawatts. Thus, the viability of tides as an energy source is greatly enhanced by the strong tides in estuaries and marginal seas around the Atlantic. Does this make tides a contender in the quest for major renewable energy sources? How much power is, in principle, available and what would be the consequences of exploiting it? The 3.2 terawatts of natural dissipation is not a direct measure of the power that is available in principle, although it turns out to be a useful guess. Extracted energy could be added to the natural dissipation, imposing an additional drag or delay that increases the angle δ in Fig. 2.2, which has a present value of 2.9°. We consider below how this angle arises. The extraction of tidal energy is fundamentally different from the atmospheric situation in which there is a fixed amount of power that is fed down from the high atmosphere and can be intercepted. Tides do not offer renewable energy in the strict sense, because they irreversibly destroy rotational energy. Tidal power generators add to the natural dissipation, not substitute for it. They would accelerate the slowing of the Earth's rotation and the recession of the Moon, although only on a multi-million year time scale.

In assessing the extent to which strong tides at the ocean margins could be harnessed, it is necessary to recognise that they are not caused directly by the tide raising gravitational forces of the Moon and Sun, but are driven by the tides in the deep oceans. The lack of significant tides in the Mediterranean makes this obvious. Tides are a global-scale phenomenon, being insignificant in small bodies of water with little or no ocean connection. Although some tidal currents have impressive kinetic energy, a simple mechanical analogue illustrates a fundamental limitation on the accessibility of this energy. Consider a weight hanging on a string and gently swing the top of the string to and fro in the manner of a simple pendulum (Fig. 14.2). When the period of the imposed motion coincides with the natural period of pendulum oscillation, it is resonant and develops a swing that is much wider than the imposed motion at the top of the string. But when the weight, or a wire protruding from it, is allowed to dip into a dish of water, energy is expended in stirring the water and the amplitude of the motion is reduced. The flow of energy into the water does not correspond to the energy of the freely swinging pendulum, which does not exist when the motion is damped. The energy transmitted to the water is limited to the energy put into the pendulum by the imposed motion at the top of the string. The same principle applies to the energy of the tide in a bay or estuary that is resonant because of its size and shape. If the water is impounded, delaying the tide, or if it is made to drive a submerged turbine, the tidal energy is reduced and no more can be extracted from it than is fed in by the open ocean tide that drives it. Tidal energy is made accessible by local shallow water resonances, but its source is the open ocean on which the gravitational tidal force operates.

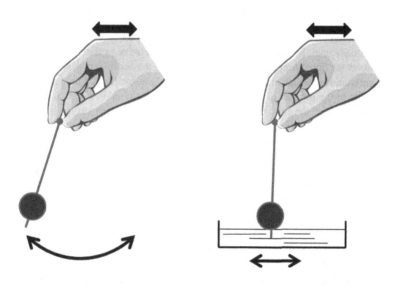

Fig. 14.2. Resonance and damping of a simple pendulum. If driven at its natural frequency it swings to a wide angle but if the motion is damped, on the right, the amplitude of the motion is reduced. No more energy can be extracted from it than is fed into it by the driving force at the top.

The tidal resonance problem is compounded by the fact that the margins of the Atlantic Ocean are particularly favourable for harnessing tidal energy because the ocean itself is close to resonant. This means that the limitation on the energy that can be extracted from a resonant system, illustrated by the pendulum analogy, applies not just to local marine resonances but to the oceans as a whole. A geometrical representation of the tidal lag in Fig. 14.3 draws attention to the quite different behaviour

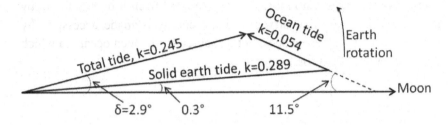

Fig. 14.3. A vector representation of tidal components (not to scale). Arrows represent the magnitudes and alignments of tidal deformations relative to the tide-producing force of the Moon (or Sun). Amplitudes are represented by the coefficient k, which is the ratio of the gravitational effect of a tide to that of the Moon (or Sun). Energy dissipation is proportional to the product of k and the lag angle, δ, as plotted in Fig. 6.3. The total tide is the vector sum of the solid earth and ocean tides. An intuitively surprising observation is that the ocean tide is inverted, in the sense of being nearly opposite to the direction of the force causing it. This is a consequence of the fact that the tide-raising force moves round the Earth faster than the natural speed of the tidal wave. Harnessing of ocean tides means modifying the phase angle marked 11.5° in the figure, but the total energy that can be extracted is represented by the resulting effect on the angle δ and amplitude, k, of the total tide.

of the solid earth and marine tides. The tidal gravitational effect (the value of k) for the solid earth tide is bigger by a factor exceeding 5, so that manipulation of the marine tide can have only a modest effect on the total tide. Harnessing the ocean tide would turn the ocean tide vector clockwise in the figure, but the consequent increase in δ would be lessened by a reduction in the amplitude of the marine tide, as represented by the length of its arrow. We can present only a qualitative estimate of the total effect, which depends on the fine details of ocean geometry, but conclude that the tidal energy extractable in principle would not exceed the natural dissipation, 3.2 terawatts, and extraction of more than a small fraction of this limit is impossible. Tidal energy may be useful in particular local situations, such as the Rance estuary at St. Malo, on the Atlantic coast of France, where a tidal power station has been producing an average power of 60 megawatts since 1966. However, although this is a particularly favourable site, it makes a miniscule contribution to global energy [14.7]. Tidal power cannot be counted as a potential major contributor to the global energy budget.

River flow

Hydroelectricity makes a significant contribution to the global electricity generation, as represented by the entry in Table 14.2, and it continues to increase. It is already more than 6% of the ultimate theoretical limit given by the river flow entry in Table 14.1, which is calculated by assuming that every drop of water that flows into the sea passes through turbines of 100% efficiency all the way from its source [14.8]. To the extent that they are available, hydroelectric facilities are ideal power sources. Gravitational energy of impounded water is stored indefinitely and can be called upon rapidly for efficient conversion to electric power. Although it is impossible for hydroelectric generation to come close to the theoretical limit of its availability, and, even by this limit, it cannot be more than a 'second tier' power source, there is great scope for its expansion in the secondary role of pumped storage facilities. At the end of this chapter we refer to its relevance to the intermittent nature of other energy sources.

Although energy derived from hydroelectric installations meets obvious requirements for our definition of 'renewable', the dams that are built for it are not without ecological and environmental problems, adding a negative footnote to the word 'renewable'. Also there is a basic difficulty of limited life times as reservoirs gradually silt up, reducing their capacities.

Waves

The energy of open ocean waves of peak-to-peak amplitude 2 metres and a period of 10 seconds propagates at 7.8 m/s, transmitting wave energy at the rate of 78 kilowatts per metre of wavefront. Allowing for refraction towards a coastline, the rate at which power could be intercepted by a near shore installation is about 40 kilowatts per metre of wavefront. This concentration of energy appears attractive from the perspective of small scale engineering, especially as it can be more or less continuous, although difficult to harness even on favourable sections of coastline. Under normal conditions some energy is reflected, but much of it is dissipated at shorelines, which globally amount to about 100,000 km, if we select only coasts with well developed waves. Using this number, we arrive at the wave dissipation estimate in Table 14.1. It is obviously unrealistic to suppose that waves could be harnessed on what would effectively be the world's entire wave-exposed coastline, or that the mechanisms for converting the wave energy would have high efficiencies. Although this estimate is very rough, it suffices to show that waves can never be more than a minor contributor to global energy.

A comment on the use of hydrogen

With currently available technology, proposals for a 'hydrogen economy' are no more than a distraction, diverting attention away from the energy problem that we face. Hydrogen occurs naturally on the Earth only transiently and in minute traces. Almost all (96%) of the commercial production is from the fossil fuels, principally natural gas, which is largely composed of methane (CH_4). The process of producing hydrogen from it discharges more CO_2 than would the combustion of the gas used in making it. Also the energy available from the hydrogen is less than in the original gas and electric power used in its manufacture, such that the resulting hydrogen offers at best a 50% return on the energy invested in it.

Hydrogen is the normal fuel for fuel cells, which make very efficient use of it by an electrochemical reaction with atmospheric oxygen in specially prepared membranes, converting the chemical energy directly to electrical energy. But fuel cells are very particular about the purity of the hydrogen. They are 'poisoned' by the normal commercial hydrogen produced from natural gas, and require hydrogen from electrolysis of water, with no more than one part per million of 'poison'. Electrolysis requires electrical power. An electric current

Edmond Becquerel, 1820–1891
One of a family of French physicists and father of Henri Becquerel, the discoverer of radioactivity. The best known of his research concerned phosphorescence and the earliest photographic experiments. In 1839, he made the first observations of the photovoltaic effect as a fortuitous discovery in experiments on electrolysis. This is the principle of modern solar cells, which use specially prepared semi-conductors. Lithograph by Pierre Petit.

passed between a pair of electrodes in water releases hydrogen at one electrode and oxygen at the other one, but pure water is a poor electrical conductor and for a useful yield a suitable solute must be added. If this is common salt (NaCl), then chlorine is liberated instead of oxygen, so that some electrolytic hydrogen is generated as a by-product in the manufacture of chlorine.

Modern electrolytic cells used for hydrogen production are quite efficient, in the sense that the energy of the hydrogen produced is about half of the electrical energy used to make it. If the hydrogen is used in fuel cells, then it effectively acts as a storage medium for electrical energy, transferring the original energy used to make it to the electrical output of the fuel cells, with only moderate losses. That is the ideal situation, but it is still less efficient (and more expensive) than storing the energy in batteries. If the hydrogen is used in the internal combustion engines of conventional vehicles, an inefficiency factor (35% or less) is introduced. If the original electrical energy came from thermal power stations, then a further factor of about 35% is applied to the overall energy efficiency. The use of hydrogen as a fuel for any but very special situations will make no environmental or economic sense until electricity generation by 'alternative' sources (solar, hydro-power, wind, etc.) becomes so abundant that the intrinsic inefficiency of hydrogen can be tolerated. Moreover, the concern over efficiency is compounded if the hydrogen is either compressed or liquefied for storage or transport because these processes consume power. If hydrogen has particular advantages as a fuel, they are its lightness [14.9] and the fact that its only combustion product is water.

Solar and wind energies are the only serious options

In recapitulating the conclusions from this brief survey of 'alternative' energy sources, we applaud efforts to harness what we regard as minor contributors, which can reduce the energy problem, but we emphasise that they cannot solve it, even in principle. The crucial conclusion is that there are only two renewable energy sources that are sufficiently abundant to supply the present energy demand, that is solar and wind. It is not incidental that they are very widely distributed being available over most of the Earth and not just in restricted locations. As comparison of the numbers in Tables 14.1 and 14.2 shows, the energy problem is of global scale and it can hardly be surprising that the solution needs to be global. But, having identified the two energy sources on which attention must be concentrated, we face the problem that both are intermittent, predictably so in the case of solar energy, as well as being unevenly distributed over the globe. A major energy storage system is needed and this is a role for hydroelectricity. If and when it is available it can be held in reserve for down time failures of both wind and solar power, but that availability is limited and development of pumped storage facilities offers greater scope. On a small scale it is an established technology in many places. Storage of water in the upper of two reservoirs is required for hours or days and not months, as for normal hydroelectric generation, but for conversion both ways, to and from electric power, with an efficiency greater than 85%, a substantial height difference is needed and this imposes geographical limitations. The down time of combined solar and wind power (windless nights) is less than for either separately, but there is a latitude bias. Wind is generally stronger and more consistent at high latitudes, where solar radiation is weak.

As a counter to often expressed pessimism about renewable energy options, we quote from a detailed study by M. Z. Jacobson and

M. A. Delucchi [14.10]: "The amount of wind power plus solar power available in likely developable locations over land outside of Antarctica worldwide to power the world for all purposes exceeds projected world power demand by more than an order of magnitude". The article concludes, not only that a plan for replacement of all fossil fuel use by 2050 is feasible, but that "Barriers to the plan are primarily social and political not technological or economic". To that we add that implementation of such a plan will be unavoidable eventually, so the sooner the better.

15. The cradle is rocking

The Earth can sustain life for many millions of years

The basic physical framework of the Earth that made life possible, the subject of our early chapters, changes only slowly, and we see no fundamental processes that would prevent the underlying habitability from continuing almost indefinitely. Solar physicists warn that the Sun will gradually heat up, but the time scale for that is billions of years. The energy source maintaining the magnetic field that protects the atmosphere from the solar wind and all but the most energetic cosmic rays is not running out; if anything it is increasing [15.1]. It may be derived in part from the precessional motion of the Earth, which is driven primarily by the Moon, and we know that the Moon is gradually receding from the Earth by tidal friction, but the time scale for that is also billions of years. As pointed out in Chapter 5 (Table 5.1), much of the thermal energy that drives plate tectonics, volcanism and the consequent environmental developments is produced by radioactive isotopes from which the energy still to be released exceeds the energy released so far in the life of the Earth. Milankovitch cycles will continue and there will be changes in the output of the Sun, but ice ages have come and gone. They were not terminal and life adjusted. Contrary to some ideas, we discount asteroidal or meteorite impacts as a major global concern and see the most serious long term environmental threat to be flood basalt volcanism, episodes of which are identified in Chapter 9 as the causes of mass extinctions of species. Although they have had an average recurrence time of about 25 million years, they have not all been equally severe. There have been two that can properly be described as devastating in the 550 million years of well developed fossils, 250 million and 65 million years ago. We can infer that, on average, one volcanic episode with the potential to wreck civilisation as we know it occurs every 200 million years or so. And they do not happen without precursory indications. Relative to the time scales of industrial environmental effects, these can be regarded as infinite times. We must think of the Earth as offering life support for an indefinitely long time, but it is finite and, with our present ability to exploit it, we are ignoring that fact and limiting the support it offers by treating it as an infinite resource and a waste dump of infinite capacity.

Fossil fuel burning is irreversibly changing the environment

As pointed out in Chapter 14, the use of energy is comparable to the energies of many global scale natural phenomena, and in Chapter 13 we see a serious change in the composition of the atmosphere. These effects emphasise the finiteness of the Earth. Environmental degradation is obvious on a local scale and is not restricted to the modern, industrial era. Deforestation began thousands of years ago and has become extensive enough to have climatic consequences. But such things are qualitatively different from what is happening now. In principle, many of them could be reversed in a few hundred years, but that is not true of the modern industrial effects, of which CO_2 emission attracts most attention. Global changes are under way and they cannot be reversed in less than many millions of years, but their visibility is reduced by natural delays arising from inertia of the system. It responds slowly, but once moving it is not readily stopped and it does not have brakes that we would be able to operate. Even if it did, would the stopping time be as long as, or even longer than, the start-up time? Consequences of the changes we have to consider take not just decades, but hundreds of years to take full effect. Remedial action will be impossible and, as we repeatedly point out, natural remediation will take millions of years.

In Chapter 13 we discuss the concept of an equilibrium state, meaning one in which the atmosphere and oceans have reached a balance with respect to both CO_2 concentrations and temperature. A crucial conclusion is that no equilibrium state is possible without completely stopping the emission of CO_2. Natural processes, aided by activities such as planting trees, can have only small incremental effects and do not significantly limit the indefinite accumulation of emitted CO_2. We are living in a transient state and it is not immediately obvious how different an equilibrium state would be. Apart from very exceptional times, such as during a flood basalt episode, the natural processes that release and sequester CO_2 act slowly. So slowly that they did not noticeably affect the total atmospheric and ocean content during the last 400,000 years, the period represented in Fig. 13.2. If and when we stop releasing CO_2, the Earth will approach equilibrium with the total emissions up to that time added to the atmosphere and oceans. The approach will take hundreds of years, as the oceans and atmosphere adjust to the increased CO_2 content and higher temperature. Neither will return to the pre-industrial level but will remain higher, effectively permanently, both because the total CO_2 has been increased and because the ocean share is reduced at the higher temperature. By 'permanently', we mean until the natural balance can be restored over millions of years. With that interpretation of the word 'permanent', all of the CO_2 that we have released will remain as a permanent change to the environment. And the global temperatures will remain higher than without the added CO_2.

Our conclusions rely, in part, on the data in Fig. 13.2, which indicate that, for at least 400,000 years before the industrial era, the total CO_2 content in the atmosphere and oceans was constant. The atmospheric CO_2 variation did not cause the temperature changes, but was caused by them, due to the temperature dependence of its solubility in the oceans. Our present situation is quite different. CO_2 is increasing in both the atmosphere and oceans (at approximately equal rates by the numbers in Chapter 13). We re-emphasise that it will not return to the pre-industrial 'normal' for millions of years and the consequences will last just as long, whatever other changes occur. Whether or not ice ages or prolonged high temperatures come and go, the industrial era greenhouse warming will be superimposed on them. Thus, we must consider two very different relaxation times, both longer than a human lifetime. One, hundreds of years and perhaps 1000 years, is the time that the atmosphere-ocean system will take to adjust to the increased total CO_2 that is imposed on it. The second one, several million years, is the time required for natural restoration of a CO_2 balance that will remove the industrial era excess.

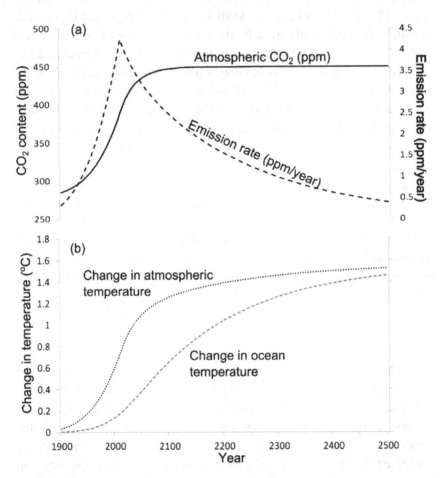

Fig. 15.1. The 450 ppm model of the greenhouse effect, which assumes that CO_2 emission is constrained so that the atmospheric concentration does not exceed 450 ppm [15.3]. (a) Variations in atmospheric concentration and emission rate, (b) Temperature rises in the atmosphere and oceans. Both are subject to extended delays as the ocean takes up heat in approaching equilibrium with the atmosphere.

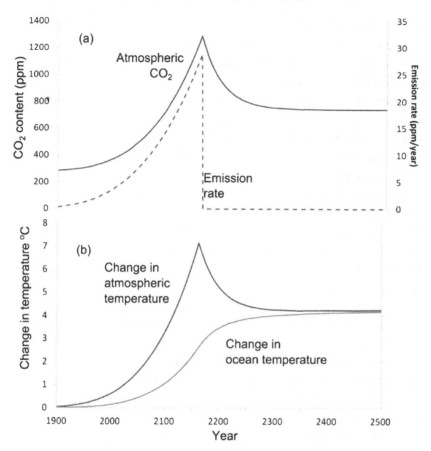

Fig. 15.2. The Business as Usual model, which assumes a simple extrapolation of the historical record of greenhouse gas emission, as in Fig. 13.1(a), until fossil fuels are exhausted in 2163. (a) Atmospheric CO_2 concentration and emission rate. (b) Atmosphere and ocean temperatures, showing relaxation towards equilibrium when emission is cut off.

The Earth is responding slowly to the rapidly changing human influence

Of the resources that are vulnerable to exhaustion, the fossil fuels, and especially oil, are most obvious. They formed over about 400 million years. Can we make them last more than 400 years, a millionth of the time taken to develop? That is a measure of the speed with which we are consuming our natural inheritance. The time that has elapsed since the civilisations of ancient Greece and Rome is much longer than the possible duration of profligate exploitation of that inheritance, which is short, even in terms of human history. It is also shorter than the time that the Earth will take to respond to greenhouse warming and puts an interesting complexion on attempts to predict what will happen. Will the international resolve to curb CO_2 emissions develop so slowly that exhaustion of the fossil fuels makes any resolve irrelevant? Whatever course is taken, the physics of the consequent processes is sufficiently well understood to allow us to outline possible alternative futures. Notwithstanding differing estimates of a greenhouse-induced temperature rise, our understanding of the Earth and of the natural controls on the environment have become very sophisticated, encompassing the diverse scientific disciplines that are needed for the complex natural system. We have the scientific basis required for decisions about human-induced environmental effects, but the seriousness and urgency of the need for attention to these effects is still generally underestimated. Although detailed modelling of the environment is a complex problem subject to ongoing research, it is possible to understand the principles with a simple, global approach that uses averaged atmosphere and ocean properties, and we apply this to two simple, extreme future scenarios, to indicate the range of possibilities. These are not complete or accurate models, but simplified ones, intended to help in understanding the detailed work of atmospheric modellers, by outlining the essential features.

Modelling atmospheric changes

For both of the models, presented in Figs. 15.1 and 15.2, up to 2010 the historical record of CO_2 is used, with the rate of solution in the ocean constrained by the observation that, in 2010, the rate of emission was twice the rate of atmospheric accumulation [7.2, 15.2]. We refer to the first model as the '450 ppm model' as it adopts the extremely optimistic aim, suggested in IPCC reports, of constraining emissions so that the atmospheric CO_2 concentration is restricted to 450 parts per million. This limit is imposed on the model by extrapolating the historical record from the 2010 situation (390 ppm increasing by 2.14 ppm/year) with no instantaneous change in the emission rate but a reversal of its rate of increase [15.3]. An exponential approach to the 450 ppm limit is assumed and the total emission includes the CO_2 dissolving in the oceans according to Equation (2) of Note [15.2]. The continuing emission at year 2500 is accommodated by solution in the oceans, which will not have reached equilibrium, even at that date, although there is negligible continuing change in the atmospheric content. The model permits a diminishing level of emission for a very long time.

The second model (Fig. 15.2) is our most pessimistic. We refer to it as the 'Business as Usual model', because it assumes that there is no constraint on CO_2 emission, which is a simple extrapolation of the expanding use of fuel in the historical record [13.1, 15.4] until the time when the total of the readily accessible fossil fuels is exhausted and emission stops suddenly. From our estimate of fuel abundances in Chapter 12, this would occur in the year 2163, far sooner than the 300 year or 400 year estimates of the time span of fuel availability at the present rate of use. After that, continued solution in the ocean reduces the atmospheric content towards the new equilibrium level of 740 ppm. The peak of this model gives an emission rate 6.8 times the 2010 rate and a total CO_2 release 8 times the 2010 total. (In 2010 one eighth of the known resource, in all its forms, had been consumed). The suddenness of the cut-off is, of course, artificial, but it does not affect the limit that is approached in the equilibrium state.

Temperature changes resulting from increases in atmospheric CO_2 are controlled by the partitioning of the increase in heat received at the surface between thermal re-radiation of heat at the higher temperature and heat retained and absorbed, primarily by the oceans. The absorption of heat delays the rise in temperature because less heat is thermally re-radiated from the surface. This is a transient state because transfer of heat into the oceans occurs only as long as there is an imbalance between the atmospheric and ocean temperatures and the rate of transfer is proportional to the imbalance. Guided by the data reported in the paper cited in Note [8.7], our models take the 2010 rate of heat transfer to the oceans as 180 terawatts. This is a measure of the departure from thermal equilibrium and is used in the model calculations outlined in Note [15.5]. An effect that is not allowed for in our models is variable cloudiness and particularly the vague, hazy cloudiness caused by fine particulates (aerosols) derived from exhausts of high-flying aircraft as well as ground-based industry. This reduces the greenhouse-induced temperature rise. Our modelling implicitly assumes that, since the variation in cloudiness is already well established, it is included in the presently observed effects and therefore also in calculations that are essentially extrapolations from present conditions.

Sea level rise will accelerate

The temperature changes plotted in parts (b) of Figs. 15.1 and 15.2 are at the low end of the range of temperature changes obtained in more detailed atmospheric models. These are compared with one another by calculations of the temperature increment for an equilibrium state with doubling of the atmospheric CO_2 content, to 560 ppm, relative to the pre-industrial level of 280 ppm. For our models this increment is 2.6° [15.7]. This is at the low end of the range of results from more detailed atmospheric models (2.1° to 4.4°), suggesting that all of our values may be biased low. There are several reasons why this may be so. An obvious one is that our models neglect the decrease in solubility of CO_2 in the oceans with rising temperature, which explains the variations in atmospheric CO_2 through the ice ages (Chapter 13). That effect depends on ocean temperature, which is delayed relative to atmospheric temperature, and so becomes more significant as equilibrium is approached. For the Business as Usual model, it raises the equilibrium CO_2 concentration to 821 ppm and the temperature increment to 5.4° (from 4.28°) [15.8]. But the solubility of CO_2 in the oceans under these conditions is not well known and, since it decreases with increasing acidity, much larger effects are possible. We therefore rate our models as lower bounds on what to expect. Temperature rises greater by a factor 1.5 are likely, making the 2163 peak of the Business as Usual model 11°C.

A well understood consequence of global warming is sea level rise and we give this some attention in Chapter 8. Although warming of the oceans accounts for most of the present net heat input to the Earth, and contributes to sea level rise, we reassert a conclusion from Chapter 8 that this is not the main cause for concern about sea level. In the long term, ice cap melting is far more significant. Arguments that follow here confirm that the sea level rise by thermal expansion is a relatively minor effect. Accepting the estimate that, in recent times, it contributed 0.8 mm/year to sea level rise [8.7], and applying the conclusion that the relaxation time for establishment of thermal equilibrium with the atmosphere (at which heating of the ocean would stop) is about 130 years [15.5], the eventual total rise that could be accounted for in this way with the present temperatures is about 130×0.8 mm = 10.4 cm. Taking the equilibrium end point of the Business as Usual model (after year 2300) as the extreme case, the total rise attributable to thermal expansion would not exceed 80 cm. But that would entail extreme heating of the Earth and ice cap melting that would cause tens of metres of rapid sea level rise. Considerations of the sea level problem can safely neglect thermal expansion and must concentrate on the processes of heat transfer to the ice caps. What is the mechanism of heat transfer and why do we expect its importance to increase?

There are two factors to take into account in approaching this question. First, we note that the present rate of melting of the north polar cap is more obvious and apparently faster than the loss of Antarctic ice, and see the asymmetry in the distribution of continents and oceans as the reason. While the oceans are absorbing almost all of the net heat input they do so more effectively in the southern hemisphere, which has more ocean. Excess atmospheric heat is most abundant at low latitudes and has a great expanse of ocean to cross to reach Antarctica, so that much of it is absorbed on the way by ocean that is below the temperature of equilibrium with the atmosphere. Conversely, the Arctic ice cap is exposed to air from the continents over a large fraction of its perimeter. When the oceans approach thermal equilibrium with the atmosphere that difference will disappear and in both hemispheres warmed summer air will more effectively reach the ice caps. Another consideration is that, in the infrared (thermal) range of radiation, the oceans are close to being ideal black bodies, as envisaged by radiation laws [3.1]. This means that they are very effective emitters, as well as absorbers, of thermal radiation, and excess atmospheric heat conducted into them is radiated away quickly if it is not stirred into deeper water. This also ceases to operate when the atmosphere and ocean are in equilibrium. Thus, although it is a simplified interpretation, we get a sense of what is happening by regarding ocean heating and ice cap melting as mutually exclusive sinks of the net greenhouse heat input. Ice cap melting will accelerate as the ocean temperature approaches equilibrium with the atmosphere. Implausibly extreme conditions would be required for ocean warming to cause a 1 m rise in sea level, whereas complete melting of the ice caps would cause a level rise that could be as much as 80 m, if it occurs rapidly, reducing to about 55 m by isostatic adjustment over several thousand years.

With this interpretation there is a sense of urgency in the threat of sea level rise, even though it does not appear so because temperature changes will occur on the time scales of Figs. 15.1 and 15.2. Effects, such as the eventual magnitude of the sea level rise that must be expected, are obscured by the inertia of the system, with the oceans absorbing both CO_2 and heat. The eventual approach to equilibrium will reveal the magnitudes of the effects that have already been set in train and are effectively irreversible because they will last for millions of years. The longer we delay recognition of this and continue living in a transient state, with the eventual outcome obscured, the more serious that outcome will be and the question of sea level rise presents a readily understood focus. The world's land area is 142 million square kilometres, approximately 2 hectares (5 acres) per person (2010 population 7 billion). This area includes mountains, glaciers, marshes and deserts. Sea level rise will inevitably accelerate, eventually engulfing much habitable and productive land. Population pressure has strongly influenced human history. Although it has been mitigated by improvements in the production of food (and almost everything else) in the industrial era, populations have increased correspondingly, and population pressure will return with a vengeance with the diminished land area. Delays built into the system do not absolve the present generation from responsibility for consequences that may be 4 or 5 generations away.

We are on a one way road

We are at the early stage of changes to the environment that will make it different in several ways from our own experience. One of the differences will be a reduction in biodiversity — and not just because of exterminations resulting directly from human activity. Evolutionary adjustment and replacement of species that do not adapt to environmental change is slow and will not keep up with the rapid changes we are causing. And those changes will be effectively permanent. Apart from any minor adjustment by sequestration, virtually all of the CO_2 we put into the atmosphere-ocean system will still be there in a million years time and the consequences of that will last just as long. Unless CO_2 release stops completely, we will inexorably approach the worst case scenario that we have referred to as Business as Usual. Our discussion uses sea level rise as an illustration of what is happening, partly because its consequences are well understood, but it is far from being the most serious environmental effect and is better described as an inconvenience than as an environmental problem. More serious than sea level rise will be the acidification of the oceans, which will probably have more dramatic effects, on land as well as at sea, than rising temperature. Are we living at a tipping point in the geological and environmental history of the Earth? To answer this question we need a careful definition of 'tipping point'. If we accept the usual sense, a point at which a return to previous conditions becomes impossible or there is an unstable acceleration of introduced changes, then, by our interpretation, it is not a relevant concept. There is no turning back from the changes already initiated by fossil fuel burning. All we can do is to minimise further changes.

16. A summary of salient conclusions

Numbers in parentheses indicate relevant chapters.

• There is no reason to suppose that the formation of the Earth and Moon, and their evolution, required any special conditions or processes (2).

• Variations in the output of the Sun and of the Earth's orbit will continue, and will cause climatic variations, but they will be within the range to which life can adjust (3).

• Geological processes, driven by the Earth's internal heat will continue for billions of years with little change, except for the threat of flood basalt volcanism every hundred million years or so (5, 9).

• The magnetic fields of the Earth and Sun, which protect the Earth from interplanetary and interstellar particles, are robust, although variations will occur with possible climatic consequences (4).

• Although the Earth has taken 4.5 billion years to reach its present state, continuing changes are extremely slow and, noting the first four points, we have the basis for life into the indefinite future (1, 5, 15).

• The oceans have several crucial environmental influences, without which life, as we know it, would be impossible. They are unique in the solar system and are essential to the evolution of the Earth to its present state (6).

• The Atlantic Ocean is close to tidal resonance and this has caused a three-fold increase in the rate of dissipation of tidal energy, globally, over the last 100 million years or so, with a corresponding increase in the rate at which the Moon is receding from the Earth. It is necessary to allow for this in discussing the history of the Moon and its influence on the Earth (2, 6).

• Environmental consequences of increasing ocean acidity, caused by solution of CO_2, may be at least as serious as the greenhouse-induced temperature rise (13, 15).

• Abundant free oxygen in the atmosphere was slow to appear and was the result of two processes, both of which are necessary. Photolysis of water vapour in the upper atmosphere, with the loss of hydrogen to space, contributed at least 60%, more or less steadily, for 4.5 billion years and was almost entirely consumed by weathering. The hydrogen loss was primarily of the light isotope, as now explained by a gradient in the deuterium/protium ratio in the stratosphere. Photosynthesis has produced about 40% of the total oxygen, but started much later. It is now the dominant source. With both sources evidently secure, atmospheric oxygen is not vulnerable and is not threatened by fossil fuel burning (7, 12).

• Natural sources and sinks of CO_2 are closely balanced, maintaining the total abundance in the atmosphere and oceans constant on a million year time scale, with temperature-controlled exchange between them. CO_2 released by fossil fuel burning is an addition that will remain in the atmosphere-ocean system for millions of years (13).

• Water vapour is a strong greenhouse gas but, being a condensing gas, it responds to temperature changes but cannot cause them. Its atmospheric concentration is controlled by temperature and increases if temperature rises, reinforcing the rise, but that is simply adding feedback to the warming caused by non-condensing gases, of which CO_2 is the most important (8).

• The greenhouse effect does not cause warming in all situations. A negative greenhouse effect (cooling) is seen in the stratosphere, where the air is tenuous, and on Mars, where the atmosphere has a high CO_2 concentration but is dry as well as tenuous (8).

• The high surface temperature of Venus cannot be attributed to a greenhouse effect of its dense CO_2 atmosphere, because very little solar radiation reaches the surface. A radiation balance is maintained high in the almost opaque atmosphere. The relationship between the temperature there and at the surface is a result of compression heating, driven by atmospheric motion. This is quite different from the terrestrial situation (8).

• The Earth's surface temperature is maintained by a combination of CO_2 and water vapour, but this greenhouse control is not instantaneous. Although increasing CO_2 causes increasing temperature, thermal inertia of the oceans slows the response and, even if emissions stopped now, the full impact of industrial era CO_2 emissions would not be felt for at least a century (8, 15).

• Sea level rise will accelerate dramatically as ocean temperatures approach equilibrium with the atmosphere and increased greenhouse heat is transferred to the polar ice caps (8, 15).

• Although absorption of heat by the oceans is temporarily concealing the magnitude of the environmental changes that have been initiated, they will be permanent, in the sense of lasting for millions of years. This is the time scale of natural remediation processes. The CO_2 added to the atmosphere-ocean system by human activity will remain in the system indefinitely (13, 15).

• An approach to a stable equilibrium state of the environment, and avoidance of an eventual extreme greenhouse effect, will be possible only with complete cessation of CO_2 emissions that are not fully compensated by sequestration (13, 15).

Notes

[1.1] Lord Kelvin, *Science* **9** (No. 228), 665–674 and (No. 229), 704–711 (1899).

[1.2] C. King, *Am. J. Science* **45**, 1–20 (1893).

[1.3] Atoms consist of physically small, positively charged nuclei surrounded by negatively charged electrons that are responsible for all chemical properties. Some nuclei are unstable and spontaneously decay to nuclei of different elements, emitting radiation as they do so. This is radioactivity. It was first recognised to occur with the very heavy elements, uranium and thorium, which appear as an isolated group at the right hand end of the nuclear energy graph in Fig. 1.1. The effect of radioactivity is to drive nuclear matter downwards in the figure and for the heavy elements this means throwing off fragments (alpha particles, which are helium nuclei, and electrons). It is useful to note that the lower an element (more specifically an isotope) appears in Fig. 1.1, the more stable it is and the most stable isotopes, which lie below the general trend of the curve, are the most abundant. This is seen by comparing the positions in the figure of helium 4, carbon 12 and oxygen 16, with their solar abundances in Table 5.3. Iron 56, which is the lowest energy state of nuclear matter, is also one of the abundant isotopes. The most common isotopes are those with masses that are multiples of 4 atomic mass units, reflecting the strong binding and stability of the ^{4}He nucleus.

[1.4] E. Rutherford and F. Soddy, *Phil. Mag.* (Series 6) **5**, 1576–1591 (1903).

[1.5] R. J. Strutt, *Proc. Roy. Soc. London* **A77**, 472–485 (1906).

[1.6] The rate of decay of an isotope is commonly represented by its half life, the time taken to decay to half of an initial abundance. For ^{238}U this is 4.468 billion years. At

4.54 billion years, the age of the Earth is 1.016 times the ^{238}U half life, so that ^{238}U was initially $2^{1.016} = 2.022$ times as abundant as at present. But 4.54 billion years is 6.45 times the 0.7038 billion year half life of ^{235}U, which was initially $2^{6.45} = 87$ times as abundant.

[1.7] For a derivation of the lead-lead isochron equation see F. D. Stacey and P. M. Davis, *Physics of the Earth, 4th ed.* Cambridge University Press (2008), Section 3.7.

[1.8] H. Y. McSween, *Meteorites and their parent planets*, 2nd ed., Cambridge University Press (1999).

[1.9] P. Spurný, J. Oberst and D. Heinlein, *Nature* **423**, 151–153 (2003).

[1.10] see F. D. Stacey and P. M. Davis, *Physics of the Earth, 4th ed.* Cambridge University Press (2008), Appendix H.

[1.11] On the basis of the nuclear physics known at the time, which indicated that nuclei with odd masses were less abundant than those with even masses, Rutherford estimated that the ^{235}U/^{238}U ratio was about 0.8 when uranium first formed. This was a remarkable guess, but it was a guess. These very heavy elements were formed in the sudden neutron-rich flash of a supernova and, as Rutherford could not have known, in such a sea of neutrons ^{235}U would have been destroyed by fission as soon as it appeared. These elements were formed by decay of even heavier short lived species and the whole process was quite complicated. For example a parent of ^{235}U is the plutonium isotope, ^{239}Pu, which decays to ^{235}U, but also fissions when exposed to neutrons and could not have been the source of ^{235}U in the supernova.

[1.12] See, for example, B. Runnegar, *J. Geol. Soc. Australia* **29**, 395–411 (1982).

[1.13] S. L. Miller and H. C. Urey, *Science* **130**, 245–251 (1959) reported an experiment in which amino acids, the building blocks of proteins, were produced by an electrical discharge in a mixture of common inorganic gases (H_2O, CH_4, NH_3, H_2, CO).

[2.1] M. Ekman, *Surveys in Geophysics* **14**, 585–617 (1993).

[2.2] F. D. Stacey and P. M. Davis, *Physics of the Earth, 4th ed.* Cambridge University Press (2008), Section 8.3.

[2.3] G. E. Williams, *Rev. Geophys.* **38**, 37–59 (2000).

[2.4] F. Tera, D.A. Papanastassiou and G. J. Wasserburg, *Earth Plan. Sci. Lett.* 22, 1–21 (1974).

[2.5] G. Ryder, *J. Geophys. Res.* 197, E4 5022 (2002).

[2.6] J. J. Papike, G. Ryder and C. K. Shearer, *Revs. Mineralogy* **36**, Chapter 5, 152–157 (1998).

[2.7] $\delta^{18}O = \{(^{18}O/^{16}O)_{sample}/(^{18}O/^{16}O)_{standard} - 1\} \times 1000$.

[2.8] P. H. Warren, E. D. Young and W. I. Newman, The lunar vapour impact paradox, *National Lunar Science Conference* (2008), p2123.

[2.9] One of the arguments for a giant impact on the Earth as a source of debris that accumulated to form the Moon is that the angular momentum of the Earth-Moon system is postulated to be anomalously large and that a massive oblique impact is needed to explain it. We can compare the rotational angular velocities of the Earth as it is now, and as it would be if the orbital angular momentum of the Moon were added, with the rotational angular velocities of the other planets, also merged with their satellites where relevant. As an assessment of what may be anomalous, these angular velocities are listed in the following table as fractions of the angular velocities that would give centrifugal effects sufficient to cancel the equatorial gravity. Noting that Venus and Mercury are omitted because tidal friction has dramatically slowed their rotations, it is evident that (Earth+Moon) angular momentum does not greatly exceed what is observed for the other planets. The angular momentum argument is not supportable.

Angular momenta of planets as fractions of critical values corresponding to zero equatorial gravity, $\omega(R^3/GM)^{1/2}$.

Earth	0.058
Earth 4.5 Ga ago	0.16
Earth+Moon	0.364
Mars	0.068
Jupiter	0.289
Jupiter+satellites	0.293
Saturn	0.373
Uranus	0.170
Neptune	0.160
Pluto	0.019
Pluto+Charon	1.2
Sun	0.004
Sun+planets	0.800

[3.1] Black body radiation, the fundamental theory of which was developed in 1900 by Max Planck (Planck's law), is also known as cavity radiation. Elementary treatments can be found in most basic Physics texts (e.g. D. Halliday and R. Resnick, *Physics for Students of Science and Engineering*, Wiley, New York). A small hole in the wall of a cavity is a black body, in the sense that all radiation falling on it enters the cavity and is absorbed; none is reflected from the hole. If the cavity is at a uniform temperature then the radiation from the hole is black body radiation, with a characteristic spectrum that depends on temperature but not on the size, shape or material of the cavity. Two features of it, incorporated in Planck's law, are relevant to the Earth's thermal balance discussion: (i) Wien's law, according to which the wavelength of the most intense radiation, λ_{max}, is related to absolute temperature, T, by a constant, $\lambda_{max}T = 2.90\times10^{-3}$ m.K, and (ii) the Stefan-Boltzmann law, by which the total radiated energy per unit area is σT^4, where $\sigma = 5.67\times10^{-8}$ W m^{-2} K^{-4}. For the numbers in Table 5.1, the solar constant (total radiation per unit area received at the distance of the Earth) is taken as 1361.6 Wm^{-2}, from the value at the 2008 solar minimum (G. Kopp and J. L. Lean, *Geophys. Res. Lett.* **38**, L01706, 2011) with adjustment for a value 0.12% higher at solar maximum.

[3.2] By the black body relationship the temperature increment, ΔT, caused by a 6.9% increase in radiation is given by

$$1.069 = (T + \Delta T)^4/T^4 = 1 + 4\ \Delta T/T, \text{ with } \Delta T \ll T$$

$$\Delta T = (0.069/4)\ T, \text{ with } T_{\text{Black body}} = 278 \text{ K (Table 8.1)},$$

so that

$$\Delta T = 4.8 \text{ K.}$$

[3.3] The orbital eccentricity is $e = (1 - b^2/a^2)^{1/2} = 0.01673$, where a, b are the semi-major and semi-minor axes of the orbital ellipse and $b = a(1 - e^2)^{1/2}$. The closest approach (perigee) is at $a(1 - e)$ and the most remote point (apogee) is at $a(1 + e)$, with the Sun at one focus, at a distance ae from the centre of the ellipse. By Kepler's second law of planetary motion, the line between the planet and the Sun sweeps out equal areas in equal times. This means that at distance r, the angular velocity of the planet about the Sun, $d\theta/dt$ at angle θ in its orbit, is such that $r^2 d\theta/dt$ is constant, the same at all θ. The time dt spent in the range $d\theta$ is proportional to r^2, but the intensity of radiation at distance r is proportional to $1/r^2$, so that the radiation received in this time interval, being proportional to intensity×time, is independent of r. This is a corollary to Kepler's second law: the radiation received by a planet is the same in all equal angular segments of its orbit.

[3.4] The lunar period in tropospheric temperature was first reported by R. C. Balling and R. S. Cerveny, *Geophys. Res. Lett.* **22**(23), 3199–3201 (1995). It is a very small effect, with an amplitude of about 0.01 K, requiring extensive averaging in both time and geography to be securely discerned. With the Earth-Sun distance oscillating by $62×10^{-6}$ of its mean value, by the inverse square law the insolation oscillates by $124×10^{-6}$ of the average at the same point in the orbit. The insolation is a maximum when the Earth is closest to the Sun and then the full moon illuminates the dark side, adding reflected light to the total sunlight received and also the infrared thermal radiation from the hot (sunlit) side of the Moon, making the total effective amplitude of the oscillation in insolation $144×10^{-6}$ of the Earth-Sun distance. By the Stefan-Boltzman law, referred to in note [3.1] the amplitude ΔT of the temperature oscillation is given by

$$1 + 144×10^{-6} = 1 + 4\Delta T/T$$

so that $\Delta T = 0.01$ K, as observed.

[3.5] A. H. Gordon, *Nature* **367**, 325–326 (1994).

[3.6] We can postulate that the mean northern hemisphere temperature, $T = 288$ K, changes by $\Delta T = 0.01$ K in response to a change, ΔH, in heat emission by human activity (between weekdays and weekends). Writing $\Delta H = f ×1.6×10^{13}$ W, that is, assuming it to be a fraction f of the global rate of energy use (which is predominantly in the northern hemisphere), comparing it with the solar energy reaching the surface of the hemisphere, $H = 5×10^{16}$ W (half of the global total), and applying the radiative black body law, we have $4\Delta T/T = \Delta H/H$, which gives $f = 0.43$. With the assumed numerical values this is a minimum estimate, but a change exceeding 40% appears implausible.

[3.7] The sunspot cycle is a symptom of the magnetic activity of the Sun. Its magnetic field reverses approximately every 11 years and sunspots are sites of very intense magnetic fields. Observations of other Sun-like stars (too distant to observe star-spots) also show a correlation between magnetic activity and luminosity. A detailed appraisal of evidence that the sunspot cycle has not always been regular, but was interrupted by the Maunder minimum was presented by J. A. Eddy, *Science* **192**, 1189–1202 (1976).

[3.8] The gravitational potential at the solar surface (the energy release per kilogram of material falling on it) is $GM/R = 1.9 \times 10^{11}$ J kg^{-1}. The output of the Sun is 3.846×10^{26} W. If infall were to account for a 5% increase it would require $0.05 \times (3.846 \times 10^{26}/1.9 \times 10^{11}) = 10^{14}$ kg s^{-1} of dust and gas. Over 10 million years (3.16×10^{14} s), such an infall would amount to 3.16×10^{28} kg, 1.6% of the mass of the Sun.

[3.9] A calculation of the mass loss by a star that starts as 100% hydrogen and is converted to 100% helium, radiating away the energy released, appeals to the energy difference between ^{1}H and ^{4}He in Fig. 1.1, 6.8×10^{14} J/kg. Converting this to mass by $E = mc^2$, we obtain a mass loss of 0.75%.

[3.10] In the range of masses comparable to the Sun, the luminosities of stars, L, that is the total radiative output, is related to mass, M, by the simple empirical relationship, $L \propto M^{3.5}$. Energy received by the Earth would vary as the inverse square of the distance, $1/R^2$, so that to compensate for the lower luminosity of a smaller Sun, R would need to vary as $M^{1.75}$. But the reduction in R would increase the solar tidal friction, which depends on M^2/R^6 [F. D. Stacey and P. M. Davis, *Physics of the Earth, 4th ed.* Cambridge University Press (2008), Section 8.3], and therefore would vary with M as $M^2/(M^{1.75})^6 = M^{-8.5}$. Under present conditions, solar tidal friction is 21% of the lunar effect and, acting alone, would cause a decrease in the Earth's rotation rate $d\omega/dt = -1.1 \times 10^{-22}$ rad s^{-2}. If we assume a primordial rotation rate $\omega = 14 \times 10^{-5}$ rad s^{-1} (twice the present rate), the total stopping time would be $14 \times 10^{-5}/1.1 \times 10^{-22}$ s $= 1.27 \times 10^{18}$ s = 40 billion years. This would be reduced to the present age of the solar system, 4.5 billion years, if the Sun's mass were smaller by the factor $(4.5/40)^{1/8.5} = 0.77$. A solar mass 20% to 25% smaller would mean that the Earth would, by now, have stopped rotating. (Note that the rate of slowing remains constant until rotation, relative to the Sun, stops completely).

[3.11] The lifetime of the Sun as a main sequence star (which it is now) is estimated to be 9 or 10 billion years, of which half has passed. With total energy proportional to M and the rate of its use (luminosity) proportional to $M^{3.5}$, the lifetime would be proportional to $M^{-2.5}$. This would be reduced to 4.5 billion years with a solar mass larger by the factor $(9/4.5)^{1/2.5} = 1.32$, that is if the Sun were 32% more massive than it is, it would have reached its end point.

[4.1] With the liquid iron outermost core, the core-mantle boundary temperature of Mercury is at least 2200 K, 1750 K above the mean surface temperature. This means an average temperature gradient $dT/dz \approx 0.0035$ K m^{-1} in the 500 km deep mantle. Taking the thermal conductivity to be $\kappa = 4$ W K^{-1}m^{-1}, as for terrestrial mantle material, and a mean area for the heat flow $A = 6 \times 10^{13}$ m^2, the conducted heat flux is $\kappa A dT/dz = 8.4 \times 10^{11}$ W, slightly more than the conducted heat flux in the core and therefore sufficient to cool the core adequately for weak convective dynamo action.

[4.2] D. Tarling, *Principles and applications of palaeomagnetism*, Chapman and Hall (1971).

[4.3] The core is less elliptical than the Earth as a whole because it is denser and its self gravitation is more effective in pulling it towards a spherical form against the centrifugal effect of rotation. The precessional torques exerted by lunar and solar gravity are proportional to the ellipticity, so that they do not suffice to keep the core in step with the precession of the Earth as a whole. Its precessional motion is maintained by coupling to the mantle and this causes internal motion within the fluid core. The effectiveness of this motion in driving the geomagnetic dynamo depends on the turbulent behaviour of the core and is subject to differing opinions, but needs to be kept in mind because adequacy of other dynamo energy sources is not clear.

[4.4] The significance of ^{10}Be concentrations was pointed out by K. G. McCracken and J. Beer, *J. Geophys. Res.* **112**, A10101 (2007). See also more recent statements in *Space Science Reviews*, DOI 10.1007/s11214-011-9843-3 and 9851-3 (2011). An indication of complications in the production and distribution of ^{10}Be, presented by W. R. Webber and P. R. Higbie, *J. Geophys. Res.* **115**, A05102 (2010) and references therein, means that, although this is an important effect, detailed interpretation requires caution.

[4.5] H. Svensmark, *Phys. Rev. Lett.* **81**, 5027–5030 (1998).

[5.1] The melting points of solids increase with pressure. For material at the base of the mantle the melting temperature is about three times the value for surface rocks.

[5.2] The general relationship for the rate of solid creep, or deformation under stress, of silicate minerals, due to J. Weertman, is written

$$d\varepsilon/dt \equiv \dot{\varepsilon} = B(\sigma/\mu)^n \exp(-gT_M/T)$$

where σ is the stress causing deformation ε, μ is the elastic modulus (rigidity), B, $n = 1$ to 3, and $g \approx 27$ are constants. The strong temperature dependence is expressed by the value of g. The exponential term in this equation is changed by a factor 10 for a 4% change in T/T_M if $T\ T/T_M \approx 0.5$.

[5.3] Active plumes occur in all tectonic situations, mid-ocean plate (Hawaii, Reunion Island), mid-continent (Zaire), Ocean ridge (Iceland) and subduction zone (Yellowstone).

[5.4] F. D. Stacey and P. M. Davis, *Physics of the Earth, 4th ed.* Cambridge University Press (2008), Section 23.2.

[6.1] The mathematical analysis of tides refers to the quantity k as the tidal potential Love number, representing it by k_2, for ellipsoidal (second degree zonal harmonic) deformation, to distinguish it from similar numbers for different deformations of the Earth. It is the ratio of the gravitational potential due to the tidal deformation of the Earth itself to the tidal potential of the deforming gravity field.

[6.2] If a wave travels at speed $v = \sqrt{(gh)}$ both ways across a channel of width w in time t, then $t = 2w/v$ and

$$h = 4w^2/gt^2.$$

For the lunar semi-diurnal tide, $t = 12.42$ hours. Expressing h in metres but w in kilometres

$$h = 2.092 \times 10^{-4}\ w^2.$$

For a tsunami with a 20 minute period

$$h = 0.283\ w^2.$$

These are the relationships plotted in Fig. 6.4.

[6.3] R. Van der Voo, *Revs. Geophys.* **28**, 167–206 (1990).

[6.4] E. C. Bullard, J. E. Everett and A. G. Smith, *Phil. Trans. Roy. Soc. Lond.* **A258**, 41–51 (1965).

[6.5] B. Fegley in *Handbook of Physical Constants, 1: Global Earth Physics*, ed. T.J. Ahrens, American Geophysical Union (1995), pp. 320–345.

[6.6] N. Clauer, S. Chaudhuri, T. Toulkeridis and G. Blanc, *Geology* **28**, 1015–1018 (2000).

[6.7] E. T. Degens and D. A. Ross, eds., *Hot brines and recent heavy metal deposits in the Red Sea.* Springer (1969).

[6.8] Considering only Na and Mg, we can imagine making 1 kg of sea salt mix from x kg of Red Sea salt, comprising $0.991785x$ kg of Na and $0.008264x$ kg of Mg, and $(1 - x)$ kg of river salt, comprising $0.591837(1 - x)$ kg of Na and $0.408168(1 - x)$ kg of Mg. Equating the ratio of total Na in this mix to the total Mg, to the ocean ratio, 8.432, yields an equation for x giving $x = 0.7553$. The ratio of the two inputs is $x/(1 - x) = 3.086$. A similar calculation for Cl/SO_4 gives a ratio of 3.513.

[6.9] The solar energy flux (solar 'constant'), 1362 W m^{-2}, delivered to the cross section of the Earth plus troposphere, $\pi \times (6.381 \times 10^6)^2$ m^2, for a year, 3.156×10^7 s, would yield 5.5×10^{24} J. The ocean heat capacity is the product of its mass, 1.4×10^{21} kg, and specific heat, 3990 J kg^{-1}, that is 5.6×10^{24} J K^{-1}.

[7.1] S. N. Williams, et al., *Geochimica et Cosmochimica Acta* **36**, 1765–1770 (1992).

[7.2] P. Friedlingstein, et al., *Nature Geoscience* **3**, 811–812 (2010).

[7.3] H. Kokubu, T. Mayeda and H. C. Urey, *Geochim. Cosmochim. Acta* **21**, 247–258 (1961).

[7.4] Urey himself commented on this on several occasions and we have also the early judgement of H. E. Suess (*Tellus* **18**, 207–211 (1966)), based on chemical considerations: "The escape from the Earth of hydrogen, produced by photolysis of water vapour, was important for the formation of an atmosphere containing free oxygen". There have been other papers drawing attention to this but not mentioning the isotopic evidence (e.g. R. T. Brinkman, *J. Geophys. Res.* **74**, 5355–5368 (1969)).

[7.5] H. D. Holland, *The chemistry of the atmosphere and oceans.* Wiley, (1978), p. 296, presented what has become the conventional argument, although H. E. Suess [7.4] had much earlier presented a strong chemical argument for the importance of hydrogen loss.

[7.6] This number is implied by the rate of hydrogen escape, as estimated by D. M. Hunten, 'Exospheres and planetary escape' in *Geophysical Monograph* **130**. American Geophysical Union (2002), pp. 191–202.

[7.7] W. J. Randel, et al. (7 authors), *J. Geophys. Res.***117**, D06303 (2012).

[7.8] With an ocean mass of 1.4×10^{21} kg, 5.8% is 8.1×10^{19} kg, of which the fraction 16/18, 7.2×10^{19} kg, is oxygen. This is 59 times the present atmospheric oxygen, which is 20.9% of the atmosphere by volume and 23.1% by mass.

[7.9] $8FeO + 2H_2O + CO_2 \rightarrow 4Fe_2O_3 + CH_4$.

[7.10] Wikipedia gives an excellent account of noctilucent clouds.

[7.11] http://www.youtube.com/watch?v=-xF2vSKINK0

[7.12] See A. B. Woodland, 'Banded iron formations' in *The Oxford Companion to the Earth* Oxford University Press (2000), pp. 61–63.

[7.13] D. Papineau, et al. (7 authors), *Nature Geoscience* **4**, 376–379 (2011).

[7.14] Visible light has a range of wavelengths from blue/violet at about 430 nm (billionths of a metre) to red at 690 nm, although some species (birds, insects, fishes) have vision extending into the shorter wavelength, ultraviolet range. Ultraviolet light (UV) is conventionally divided into three wavelength ranges, identified as A, B, C, according to its biological effects. UV-A has wavelengths extending from the visible range to 315 nm. The atmosphere is reasonably transparent to it and even absorption by ozone is weak over much of this range. UV-B, extending from 315 to 280 nm, which is identified as biologically damaging, is partially blocked by stratospheric ozone. Wavelengths shorter than 280 nm, identified as UV-C, are biologically very damaging and are used as a germicide, but virtually none is transmitted through the upper atmosphere. The UV absorption by sea water varies considerably between water samples. For UV-A, it is generally no more effective than is the atmosphere, but increases through the UV-B range, although less strongly than for ozone, and is most effective in the UV-C range. Absorption in all ranges is strongly increased by suspended fine particles, which would have been more abundant in the oceans before land life reduced the erosion of continents as wind-blown dust as well as river sediment.

[8.1] The solar radiation energy per unit area received by the Earth at its average distance from the Sun is referred to as the solar constant, although now known not to be perfectly constant. Its average value is $S = 1361.6$ W m^{-2}. For the Earth, radius a, and cross sectional area πa^2, the total energy received is $\pi a^2 S$. If this is re-radiated from the total surface area, $4\pi a^2$, as black body thermal radiation at temperature T, by Stefan's law it is equal to $4\pi a^2 \sigma T^4$, where $\sigma = 5.670 \times 10^{-8}$ W m^{-2} K^{-4} is the Stefan-Boltzman constant, as in Note [3.1]:

$$\pi a^2 S = 4\pi a^2 \sigma T^4$$
$$T = (S/4\sigma)^{\frac{1}{4}} = 278 \text{ K}$$

[8.2] For a review of the role of water vapour in the control of atmospheric temperature see I. M. Held and B. J. Soden, *Ann. Rev. Energy Environ.* **25**, 441-475 (2000).

[8.3] S. R. Weart, *The discovery of global warming.* Harvard University Press (2008).

[8.4] The machinations of a few politically well connected scientists, supported by industry lobby groups, who bad mouth the science and scientists documenting evidence on matters of public concern are given detailed exposure by N. Oreskes and E. M. Conway, *Merchants of doubt: how a handful of scientists obscured the truth on issues from tobacco smoke to global warming.* New York, Bloomsbury, 2010 (briefly summarised in http://www.merchantsofdoubt.org).

[8.5] D. R. Easterling, et al. (11authors), *Science* **277**, 364–367 (1997).

[8.6] A well known kitchen sink experiment to illustrate that sea level is not affected by melting of floating ice needs a little patience to be convincing. Place a piece of ice, as large as convenient, in a glass of cold water and mark the water level. Check that the level has not changed when the ice has melted, but if the experiment is speeded up by using hot water, its thermal contraction will confuse the result. Hence the need for patience in conducting the experiment.

[8.7] J. A. Church, et al. (10 authors), *Geophys. Res. Lett.* **38**, L18601 (2011).

[8.8] A 120 m rise over the ocean area of 3.62×10^{14} m^2 means an increase of 4.34×10^{16} m^3 in ocean volume, so that 53% is 2.3×10^{16} m^3. This is 1.65% of the total ocean volume, 1.4×10^{18} m^3. This is the volume increase of cold water heated to 60°C.

[8.9] N. P. McKay, J. T. Overpeck and B. L. Otto–Bliesner, *Geophys. Res. Lett.* **38**, L14605 (2011).

[8.10] In the ocean situation thermal expansion coefficient, α, is very variable because of its strong temperature dependence (Fig. 8.6) and the wide range in temperature (Fig. 6.9). By taking an average for the volume, V (mass m), that is heated, we do not need to know what that volume is. If the temperature increase is ΔT, the volume increase is $\Delta V = V\alpha\Delta T$. If this is caused by heat input $\Delta Q = mC_P\Delta T$, the ratio

$$\Delta V/\Delta Q = V\alpha\Delta T/ mC_P\Delta T = \alpha/\rho C_P = 5.57 \times 10^{-11} \text{ m}^3/\text{J}$$

where density, $\rho = 1025$ kg m^{-3} and specific heat $C_P = 3990$ J kg^{-1} K^{-1}, so that $\alpha = 2.3 \times 10^{-4}$ K^{-1}, which corresponds to 23°C in Fig. 8.6.

[8.11] The 0.8 mm/year sea level rise attributed to thermal expansion is identified with heat input of 5.2×10^{21} J/year. The latent heat of melting of ice is 3.35×10^5 J kg^{-1}, so that much heat would melt 1.55×10^{16} kg of ice per year, producing 1.55×10^{13} m^3 of (fresh) water. Spread over the ocean surface area, 3.62×10^{14} m^2, this corresponds to a sea level rise of 45 mm.

[8.12] R. Kwok, et al. (6 authors), *J. Geophys. Res.* **114**, C07005 (2009).

[9.1] F. M. Gradstein, et al., *A geologic time scale 2004.* Cambridge University Press (2005). (www.stratigraphy.org/gts.htm).

[9.2] L. W. Alvarez, et al., *Science* **208**, 1095–1108, 1980.

[9.3] V. Courtillot, *Evolutionary catastrophes: the science of mass extinction*. Cambridge University Press (1999).

[9.4] A Robock and C. Oppenheimer (eds.) Volcanism and the Earth's atmosphere. *Geophysical monograph* **139**. American Geophysical Union (2003). This is a collection of papers devoted to the effects of eruptions on the atmosphere and climate.

[9.5] As a general observation, we consider that the significance of impacts to the history of the Earth has been seriously over-emphasised. The evidence for flood basalt volcanism as the cause of mass extinctions assembled by Courtillot [9.3] is convincing. This does not imply an absence of impacts but makes it difficult to argue that they have been responsible for any mass extinctions. Similarly, the lunar orbital evolution measurements by Williams [2.3] extrapolate back to an inferred origin of the Moon at a distance of 20 to 30 times the radius of the Earth and not 3 or 4 times, as required by a popular supposition that it accreted from debris thrown up by a massive impact on the Earth. We agree with Williams's conclusion: "This tidal scenario suggests that a close approach of the Moon has not occurred at any time in Earth history".

[10.1] The Earth's rotational angular momentum is

$$C\omega = 5.86 \times 10^{33} \text{ kg m}^2 \text{ s}^{-1}$$

(Axial moment of inertia, $C = 8.036 \times 10^{37}$ kg m^2; angular rotation speed, $\omega = 7.292 \times 10^{-5}$ rad s^{-1}).

Orbital angular momentum is

$$MR^2\Omega = 2.66 \times 10^{40} \text{ kg m}^2 \text{ s}^{-1}$$

(Earth mass, $M = 5.973 \times 10^{24}$ kg; distance from Sun, $R = 1.496 \times 10^{11}$ m; orbital angular speed, $\Omega = 360°/\text{year} = 1.991 \times 10^{-7}$ rad s^{-1}).

[10.2] The average strength of the Earth's field at the surface is 5×10^{-5} T and, since it falls off as $1/r^3$ with distance r, at the boundary of the magnetosphere (10 Earth radii), it is $B = 5 \times 10^{-8}$ T. At this distance it matches the pressure of the solar wind and heliospheric field, which is $P = B^2/2\mu_0 \approx 1 \times 10^{-9}$ Pa. This pressure exerts a force on the magnetosphere which is not precisely radial from the Sun because of the Sun's rotation. At the solar surface the field is locked to the Sun and rotates with it, giving a field component tangential to the surface that is a fraction of the total field equal to the ratio of the rotational velocity of the surface (2 km/s) to the solar wind speed (500 km/s), so we estimate the component of the pressure on the magnetosphere in the direction of the Earth's orbital motion to be 1×10^{-9} Pa/250 = 4×10^{-12} Pa. The area of the magnetosphere exposed to this pressure is 1.3×10^{16} m^2, giving a force of 5.2×10^4 N and a torque of 7.8×10^{15} Nm on the 1.5×10^{11} m orbital radius. This increases the orbital angular momentum by $d(MR^2\Omega)/dt = 0.5MR\Omega dR/dt$ (using Kepler's third law by which $d\Omega/dt = -(3\Omega/2R)dR/dt$), where M is the Earth's mass, Ω is the orbital angular velocity and R is the orbital radius. The resulting rate of increase in R is 8.75×10^{-14} m s^{-1} = 2.8 km per billion years.

[10.3] A comprehensive treatment of the Yarkovsky and related effects is given by W. F. Bottke, D. Vokroulický, D. P. Rubincam and D. Nesvorný, *Ann. Rev. Earth Plan. Sci.* **34**, 157-191 (2005) and there is a brief discussion in F. D. Stacey and P. M. Davis, *Physics of the Earth, 4th ed.* Cambridge University Press (2008), Section 1.9.

[10.4] The solar radiation received by the Earth, 1361 W m^{-2} over area 1.3×10^{14} m^2, is $d\varepsilon/dt = 1.7 \times 10^{17}$ W and the radiation pressure gives a force $(d\varepsilon/dt)/c = 5.8 \times 10^8$ N, where $c = 3 \times 10^8$ m s^{-1} is the speed of light. With the unrealistically extreme assumption that this force acts in the direction of orbital motion, by the same algebra as in [10.2] the orbital radius would increase by 31,000 km per billion years.

[10.5] With solar power per unit area, $S = 1361$ Wm^{-2} and $\sigma = 5.67 \times 10^{-8}$ W m^{-2} K^{-4} the present average black body temperature is $T = (S/4\sigma)^{1/4} = 278$ K.
With an early Sun of luminosity $0.7S$, $T = (0.7S/4\sigma)^{1/4} = 255$ K.

[11.1] Uranium reactors rely on the isotope ^{235}U, which is now only 0.72% of total uranium, but was 4% two billion years ago. Its present abundance does not suffice for development of natural reactors similar to those that once operated at Oklo.

[11.2] Although diamonds are remarkably robust and can withstand extreme stresses and high temperatures, the crystal structure is the equilibrium form of carbon only at pressures exceeding about 5 GPa (50,000 times atmospheric pressure). At lower pressures the equilibrium form is graphite.

[12.1] In a molecule of CO_2, two atoms of oxygen, mass $2\times16 = 32$ atomic mass units (amu), are coupled to one atom of carbon, mass 12 amu, so release of 1.22×10^{18} kg of O_2 means extraction of $1.22\times10^{18}\times12/32$ kg of C from CO_2. Spread over an area of 5.1×10^{14} m^2 this is $(1.22\times10^{18}\times12/32)/ 5.1\times10^{14} = 897$ kg m^{-2}. Taking the average density of coal, with 70% of carbon by mass, as 1400 kg m^{-3}, if all of the present atmospheric oxygen were extracted from CO_2, depositing the carbon as coal, it would amount to a global layer of thickness 0.9 m.

[12.2] The ratio of the CO_2 released by burning fossil fuels to the thermal energy produced varies systematically with the proportion of hydrogen to carbon. Methane has the highest proportion of hydrogen and yields more water than carbon dioxide when it is burned. At the other extreme, the highest grade of coal, anthracite, is almost pure carbon, so that CO_2 is almost the only product of combustion. The following is a brief league table of CO_2 released per unit of thermal energy:

Fuel	Kilograms of CO_2 per gigajoule of energy
Gas (methane)	53
Oil	70
Coal	88
Peat	106

The unfavourable position of peat in this table is a consequence of the fact that it is the lowest grade (least processed) of the fuels and includes materials such as water, expulsion of which consumes much of the energy of combustion.

[12.3] H. D. Klemme and G. F. Ulmishek, *Am. Assoc. Petrol. Geol. Bull.* **75**(12), 1809–1851 (1991).

[12.4] D. H. Tarling, *Palaeomagnetism: principles and applications in geology, geophysics and archaeology* Chapman and Hall (1983).

[12.5] Methane emissions are reviewed by G. Etiope and R. W. Klusman, *Chemosphere* **49**, 777–789 (2002), and K. F. Kroeger, R. diPrimio and B. Horsfield, *Earth Sci. Rev.***107**, 423–442 (2011).

[12.6] cyano- is a prefix meaning blue colouration. Its use is extended to refer to compounds of carbon and nitrogen (CN), as in cyanide.

[12.7] D. J. DesMarais, *Rev. Mineral. Geochem.* **43**, 555–578, (2001), gives estimates of both reduced (organic) and oxidised (carbonate) carbon in various components of the crust. The totals are 1.6×10^{19} kg of reduced carbon, in which we include 10^{18} kg of carbon in clathrates, and 7.0×10^{19} kg of oxidised carbon. Although both values may err on the low side, they are consistent with other estimates and we use them in our calculations. If this were all released to the atmosphere as carbon dioxide, it would amount to 60 times the mass of the present atmosphere.

[12.8] S. L. Morford, B. Z. Houlton and R. A. Dahlgren, *Nature* **477**, 78–81 (2011).

[13.1] Of the empirical functions that can be fitted to the Mauna Loa CO_2 record (Fig. 13.1(a)), a good fit is obtained with a very simple one that allows an analytical approach to the problem of solution in the ocean in Chapter 15:

$$CO_2 \text{ (ppm by volume)} = 280 + 6.075 \times 10^{-8} (\text{year} - 1804)^4$$

This is the function plotted in Fig. 13.1(b).

[13.2] M. I. Budyko, *The Earth's climate: past and future.* Academic Press, New York, (1982).

[13.3] carbonic acid, H_2CO_3
    calcium bicarbonate, $Ca(HCO_3)_2$
    calcium carbonate, $CaCO_3$

[13.4] If, in equilibrium at a reference temperature T the masses of CO_2 in the atmosphere and oceans are q_0 and Q_0 respectively, and solubility in ocean water decreases by 4.4% per degree temperature rise (I. I. Gordon and L. B. Jones, *Marine Chemistry* **1**, 317–322 (1973)), then at a temperature $(T + n°)$, the equilibrium contents become q and Q, related by $q/Q = q_0/Q_0 (1.044)^n$. With constant total, $q + Q = q_0 + Q_0$, and $q \ll Q$, the fractional change in Q is much smaller than the fractional change in q, with the result that $q/q_0 \approx (1.044)^n$. The ice age ratio, $q/q_0 = 1.4$ requires $n = 7.8$, that is a $7.8°$ temperature difference between maxima and minima.

[13.5] A scientific definition of renewable energy is energy the use of which results in no change of entropy. In practical terms this means that it leaves the Earth unaltered. From the perspective of energy and temperature, solar energy and wind energy satisfy this definition because they arrive at the surface of the Earth and, in the absence of human intervention, produce heat. If we intercept and use some of this energy then there may be a delay, but it still ends up as heat.

[13.6] See M. Halbwachs, et al., 'Degassing the "Killer Lakes", Nyos and Manoun, Cameroon', in *EOS, Transactions American Geophysical Union* **85**, No 30, 27 July 2004, pp. 281 and 285.

[14.1] Taking the solar energy received at the surface as 9.8×10^{16} W from Table 14.1 and the Earth's residual stored heat as 13.3×10^{30} J from Table 5.1, the ratio is $13.3 \times 10^{30}/9.8 \times 10^{16} = 1.36 \times 10^{14}$ s = 4.3 million years. If, instead, we consider the total energy of the Earth's formation, 217×10^{30} J, the ratio is 70 million years.

[14.2] The theoretical limiting efficiency for conversion of heat to mechanical or electrical energy, that is the ratio of energy produced to the heat input, known as Carnot's theorem, is $(1 - T_{out}/T_{in})$, where T_{in} is the temperature of the source of heat and T_{out} is the temperature at which heat is exhausted. This limit can be approached only under special conditions, which include a requirement that the two temperatures are adiabatically related for material that is thermodynamically cycled between them.

[14.3] The mass of 1 km^3 of average granite is 2.67×10^{12} kg and its specific heat is 830 J kg^{-1} K^{-1}, so the heat capacity is 2.2×10^{15} J K^{-1} With cooling by 100°, the heat release is 2.2×10^{17} J. If used with 20% efficiency the energy generated is 4.4×10^{16} J = 1400 megawatt-years (1 megawatt-year = 3.16×10^{13} J).

[14.4] Data from W, E, Carter in *The encyclopedia of solid earth geophysics*, ed. D. E. James, VanNostrand-Rheinhold (1989) pp. 231–239.

[14.5] F. D. Stacey and P. M. Davis, *Physics of the Earth, 4th ed.* Cambridge University Press (2008), Section 26.2.

[14.6] C. L. Archer and M. Z. Jacobson, *J. Geophys. Res.* **110**, D12110 (2005).

[14.7] 200,000 power stations equivalent to that at St, Malo would be required to produce power equal to the global total energy demand.

[14.8] Of the world's rainfall on land, $\sim 10^{15}$ m^3 /year, roughly 25% flows to the sea in rivers and streams. This means a mass flow $dm/dt = 7.9 \times 10^8$ kg s^{-1}. By assuming this to flow from the average height of all land, $h = 840$ m, under gravity $g = 9.8$ m s^{-2}, $dE/dt = ghdm/dt = 6.5 \times 10^{12}$ W.

[14.9] In terms of energy per kilogram, hydrogen is outstandingly the lightest fuel, giving 143 MJ/kg, compared with 43 MJ/kg for jet fuel. However, if it is stored as either liquid (at a very low temperature) or compressed gas, the containers may be heavy enough to outweigh the advantage of lightness.

[14.10] M. Z. Jacobson and M. A. Delucchi, *Energy Policy* **39**, 1154–1169 (2011).

[15.1] The intuitively surprising observation of a gradual progressive increase in the strength of the geomagnetic field, as suggested by paleomagnetic data, can be understood as a consequence of the manner in which the solid inner core grows by freezing of outer core liquid, as the Earth cools. The energy released by this process is the dominant driver of the geomagnetic dynamo. With the pressure dependent melting point of core material, the rate of increase in the inner core radius, r, is controlled by the rate of cooling, but the energy release depends on the increase in inner core volume and therefore on the rate at which material is added over its surface area, $4\pi r^2$. Since this increases faster than r, the rate of energy release increases as the inner core grows. Relevant equations are presented in F. D. Stacey and P. M. Davis, *Physics of the Earth, 4th ed.* Cambridge University Press (2008), Section 22.7.

[15.2] Using the equation in Note [13.1] to represent the increment in atmospheric CO$_2$ concentration, relative to the pre-industrial 'background' level

$$C_A = 6.075 \times 10^{-8}\, t^4 \tag{1}$$

where t is the time in years after 1804 and C_A is in ppm, and writing a corresponding quantity, C_S, for the increase in ocean concentration, so that the rate of solution is proportional to the difference, we have

$$f\, dC_S /dt = A(C_A - C_S) \tag{2}$$

where f is the factor by which the effective ocean capacity exceeds the atmospheric capacity and A is a coefficient representing the rate of solution. Taking f and A as constants, with C_A a known function of t by Equation (1), we solve Equation (2) for C_S

$$C_S = 6.075 \times 10^{-8} \{t^4 - 4(f/A) t^3 + 12(f/A)^2 t^2 - 24(f/A)^3 t$$
$$+ 24(f/A)^4 [1 - \exp(-At/f)]\} \tag{3}$$

In 2010 the total rate of emission, $(dC_A/dt + f dC_S/dt)$, was twice the rate of increase in atmospheric concentration, dC_A/dt, that is

$$f dC_S/dt = dC_A/dt \tag{4}$$

at $t = 206$ years (2010 − 1804). This gives a relationship between f and A. For all plausible values of f, A differs little from the value we chose for our models, $A = 0.023$ for $f = 5$. For two reasons we use a value of f much lower than the recognised ocean/atmosphere abundance ratio. Mixing into the deep ocean occurs only over a period of 1000 years or so, longer than the time covered by our calculations, and increasing acidity caused by CO_2 solution inhibits further solution, reducing the effective value of f. But, in spite of uncertainty over f, the models are constrained by Equation (4), for which the 2010 value, 2.14 ppm/year, was well observed.

[15.3] By writing the atmospheric CO_2 content as

$$C = 390 + 60[1 - \exp(-t/28)]$$

With $t = (\text{date} - 2010)$, C approaches a limit of 450 ppm with $dC/dt = 2.14$ ppm/year in 2010. This matches the observed value, so that no sudden decrease in emission is assumed, but the trend is reversed.

[15.4] Until the year 2163, the 'Business as Usual model' extrapolates the total CO_2 emission by $(C_A + f C_S)$, following Equations (1) and (3) in Note [15.2], with the rate of solution given by Equation (2).

[15.5] The greenhouse heat balance equation is

$$GC_A = \lambda T_A + Q \tag{1}$$

where T_A is the surface temperature increment arising from an increment C_A in CO_2 concentration and Q is the heat retained in the Earth. GC_A is the increase in surface heating by the greater atmospheric opacity and λT_A is the increase in thermal radiation from the surface caused by the higher temperature. G and λ are coefficients to be estimated and in our models are assumed to be constants, an approximation to the real Earth that greatly simplifies the problem. Using observed 2010 values, $Q = 1.8 \times 10^{14}$ W

[8.7] at $T = 0.75°$ and $C_A = 110$ ppm, we have a relationship connecting G and λ, so that an independent estimate of either fixes Equation (1). A simple, if approximate, estimate of λ is based on the assumption that, of the radiation reaching the surface (9.6×10^{16} W), 50% is absorbed and then re-emitted as thermal radiation, most of the other 50% being reflected. This means that, at the average surface temperature, $T = 288$ K, the Earth emits 4.8×10^{16} W as thermal radiation. By the black body law [3.1] at a temperature higher by $\Delta T = 1°$ the radiation would be increased by $4\Delta T/T \times 4.8 \times 10^{16}$ W = 6.67×10^{14} W. Thus we can write the global value of λ (the extra radiated heat per degree temperature rise) as $\lambda = 6.67 \times 10^{14}$ W K^{-1}. Referred to unit area of the surface this is $\lambda = 1.3$ Wm^{-2}K^{-1}, which is the usual way of expressing the quantity known as the climate feedback parameter. Independent estimates vary widely, but our value agrees well with an estimate based on satellite observations of terrestrial radiation [15.6]. Using our global value of λ, we have $G = 6.18 \times 10^{12}$ W per ppm of CO_2.

To account for the absorption of heat by the oceans, we introduce a notional ocean temperature increment, T_S, which would be equal to T_A at equilibrium and which increases at a rate

$$dT_S/dt = B(T_A - T_S), \tag{2}$$

noting that this is related to Q by an effective heat capacity, H

$$Q = H \, dT_S/dt \tag{3}$$

Thus, we need also estimates of the coefficients, B and H, which, like G and λ, we assume to be constants, although recognising that in the real Earth they are not, because absorption of heat by the ocean surface may be rapid but its redistribution in the deep ocean is slow and is not a simple one stage process. Parameters B and H are less well constrained by observations than G and λ. Ocean heating is far from uniform and for the present purpose the effective ocean heat capacity is much less than would apply to uniform heating (5.59×10^{24} J K^{-1}). From the ocean temperature distribution in Fig. 6.9 we favour 24% of this, giving $H = 1.33 \times 10^{24}$ J K^{-1}, but that value is relevant for dT_S/dt in degrees per second and, since we are using years as the unit of time, it is convenient to divide by the number of seconds per year and use $H = 4.2 \times 10^{16}$ JK^{-1} for the purpose of our model calculations. Then $B = 7.65 \times 10^{-3}$ year^{-1}, corresponding to a 130-year relaxation time.

[15.6] D. M. Murphy, et al. (6 authors), *J. Geophys. Res.* **114**, D17107 (2009).

[15.7] Equilibrium means $Q = 0$ in Equation (1) of Note [15.4], so that $T_A = GC_A/\lambda$. With doubling of CO_2, $C_A = 280$ ppm and, using the global values of G and λ in Note [15.5], $T_A = 2.6°$.

[15.8] With $\Delta T = GC_A/\lambda$ and C_A increased by the factor $1.044^{\Delta T}$ [13.4], from the limiting Business as Usual model value of $(740 - 280) = 460$ ppm, the equilibrium values become $\Delta T = 5.4°$, $C_A = 579$ ppm (total atmospheric content 859 ppm).

Name index

Subject index

About the authors

Frank Stacey studied Physics at Queen Mary College, London, and took appointments in Vancouver (UBC), Canberra (ANU) and Cambridge before joining the University of Queensland. As Professor of Applied Physics he taught geophysics at several levels and is most widely known for his textbook 'Physics of the Earth', which is now in its fourth edition. After retirement from the university Frank joined CSIRO to continue research on fundamental problems, especially the thermodynamics of Earth materials at high pressure.

Jane Hodgkinson turned to a career in geology after studying at Birkbeck College, London, during her first career in banking and the commodity markets. After completing her PhD in Brisbane (QUT), Jane joined CSIRO where she has worked with climate scientists and mining companies on climate change adaptation projects and assessing CO_2 geostorage potential. Jane has also been involved in new mineral exploration techniques, remote geomorphological analysis and localised influences of tectonic events.